A Practical Guide for
Students and Officers,
Insurance Investigators, Loss
Adjusters and Police Officers

FIRE INVESTIGATION

Michael F. Dennett,

M.I.Fire E., M.I.A.A.I.

Divisional Officer
Greater Manchester Fire Service
England

PERGAMON PRESS
Oxford New York Toronto
Sydney Paris Frankfurt

U.K.	Pergamon Press Ltd., Headington Hill Hall, Oxford OX3 0BW, England
U.S.A.	Pergamon Press Inc., Maxwell House, Fairview Park, Elmsford, New York 10523, U.S.A.
CANADA	Pergamon of Canada, Suite 104, 150 Consumers Road, Willowdale, Ontario M2J 1P9, Canada
AUSTRALIA	Pergamon Press (Aust.) Pty. Ltd., P.O. Box 544, Potts Point, N.S.W. 2011, Australia
FRANCE	Pergamon Press SARL, 24 rue des Ecoles, 75240 Paris, Cedex 05, France
FEDERAL REPUBLIC OF GERMANY	Pergamon Press GmbH, 6242 Kronberg-Taunus, Pferdstrasse 1, Federal Republic of Germany

First edition 1980

British Library Cataloguing in Publication Data

Dennett, Michael F
Fire investigation. - (Pergamon international library).
1. Fire investigation
I. Title
614.8'4 TH9180 79-42763

ISBN 0-08-024741-5 Hardcover
ISBN 0-08-024742-3 Flexicover

Printed and bound in Great Britain by
William Clowes (Beccles) Limited, Beccles and London

Contents

Introduction

FIRE investigation as described in this book deals with the subject at its most basic level. No special knowledge of chemical analysis or expertise in the use of flammable gas detectors or other specialist equipment is required. Rather it is aimed at the new investigator who might use the work as initial study material, although it might be used as an *aide-mémoire* by the more experienced investigator.

Although arson is mentioned in the following pages, it should not be given undue prominence. The objective is to ascertain the cause of fire. The fact that the fire may have been started by a malicious act is important, but it is not in itself a full description of the cause of a fire.

To use the British Fire Service definition of what constitutes a statement of cause of fire, three separate items need to be decided, viz.

(1) the source of ignition;
(2) the material first ignited; and
(3) the defect, act or omission which led to the source of ignition and the material first ignited creating a fire.

The object of carrying out an investigation is to ascertain each of these three items. The investigator should therefore consider, examine and discount or accept, a malicious act in much the same way as he would consider the sun's rays or an electrical short circuit.

In most countries, fires which involve damage to property—whether a building, a form of transport or crops—result in an investigation being carried out.

The person who carries out the investigation may be from a variety of different disciplines depending upon the country or region in question. He may be the fire officer who attends the fire initially, or a specialist fire officer responsible solely for fire investigation; he may be the local policeman, or a police specialist who only deals with cases of arson. He may not be a public official at all, but a forensic scientist, a private investigator or an insurance investigator.

It may be that, instead of finding a specific cause of fire, it is sufficient for the investigator's purpose to ascertain what could not have been the cause. (On occasion, an insurance company will only pay out after being assured that the fire was not caused by malicious ignition, or even negligence, on the part of the insured.)

On the other hand, the investigator might have not only to find out the cause of the fire, but in the case of arson he may also be responsible for the search for, and arrest of, the suspect and for the preparation of the subsequent court case.

Whoever the investigator is, and however detailed the result he is trying to achieve, he must follow a predetermined course of action, if he is to complete the investigation in the most efficient manner possible. This course of action must, however, be flexible enough to take advantage of circumstances as they arise.

For instance, the order in which the collection of physical evidence, the taking of samples and the interviewing of witnesses takes place will depend, amongst other things, on the time taken from the start of the fire to the investigator getting to the scene.

If the investigator is on the scene while the fire is still in progress, he should be able to observe at first hand the spread of the fire, the colour of the smoke, any explosions, etc. He will also be able to interview witnesses who arrived at the fire before the first firemen; and, when the fire is under control, he will be able to interview the firemen who first entered the building. He will also be in a position to examine the remains of the premises before any turning over or clearing away takes place.

Alternatively, if the fire has been out for some time when the investigation starts, the investigator may be presented with a fully cleared site where the fire damaged building once stood; and he will then face the task of painstakingly searching for witnesses before he can start his investigation proper.

In all types of incidents, the investigator must aim to marry the physical evidence with the statements of witnesses and others. Where the two do not coincide, he must re-examine the physical evidence before reaching a conclusion as to which is correct, the statement or the investigator's interpretation of the physical evidence.

He must also remember that physical evidence is formed by the physical circumstances which appertain at the time of the fire and that these circumstances can have different values in different fires. They will vary as the structure varies, as the internal layout varies and as the wind speed and direction vary. The method and direction of attack by firefighters, where the jets were directed, and at what stage the fire was ventilated, will also have an effect on the physical evidence which remains after the fire has been extinguished.

If, after reaching a conclusion about any fire, the investigator feels that he could confidently justify his opinions in a law court under cross-examination, then he can rest assured that he has completed his task satisfactorily.

In the following chapters, the search for and compilation of the three items which constitute an acceptable cause of fire are dealt with more fully, together with sections on the collection of evidence, explosions, and the reasons for arson, etc. However, this is a basic guide and the only way in which the investigator will progress and increase his knowledge is by carrying out practical investigations in as many different situations as possible, by keeping an open mind, and by calling upon specialists for advice whenever he is presented with a situation outside his normal scope of operations. In some cases, the investigator can gain a lot of knowledge and collect a considerable amount of useful evidence simply by acting as the co-ordinator or information collection-point of a team of specialists in different fields such as engineers, chemists, etc. When acting in this manner, it is imperative that professional jealousy and personal ambitions are kept in check and made subservient to the need for all to work towards the common goal of determining the cause of fire.

Chapter One *Sources of Ignition and Materials First Ignited*

ALL COMBUSTIBLE substances need to be at a certain minimum temperature before they will ignite. The amount of heat required to raise a substance to its ignition temperature depends upon the type of substance, its physical state, its particle size, and whether or not a naked or open flame is applied to it.

Thus, depending on the above, combustible substances can have three temperature thresholds which may be defined as follows:

(1) Flash-Point Temperature

The flash-point temperature is reached when a substance is heated to the point where a limited amount of flammable vapour is produced. This temperature can be above or below 0°C (32°F), but whenever it is reached, any flame, regardless of its temperature, will cause momentary ignition of the vapour. The resultant flash of fire will not be maintained. This temperature is frequently used as an indication of the degree of hazard presented by a substance.

(2) Fire-Point Temperature
This is a higher temperature than the flash-point temperature and when it is reached sufficient flammable vapour is released for ignition to occur if a naked or open flame, similar to that mentioned above, is applied to the vapour. The difference is that as the vapour burns sufficient heat is produced to ensure continued release of vapour in a large enough quantity to allow combustion to continue.

(3) Self-Ignition Temperature
Unlike the first two, when a substance reaches this temperature, it will burst into flame spontaneously (without the need to apply a naked or open flame).

It will be realized from the aforementioned that sources of ignition can be in varying guises, depending upon the type and quantity of heat available to a particular substance.

The importance of a naked or open flame to the investigator is obvious, as it has the ability firstly to heat the substance and thereby liberate vapour, and then to provide the catalytic effect of the flame itself to ensure early ignition. However, as central heating and air-conditioning systems are installed at work and at home, and since many new developments do not have facilities for gas- or oil-burning equipment, flame as a source of ignition is becoming comparatively rare. Even in premises where gas- or oil-burning equipment is being installed, there are generally two levels of protection incorporated. The appliance itself has built-in safety devices; and, in case these fail, the appliance is separated from the remainder of the premises by non-combustible fire-resisting construction. The exception to the trend towards a lack of naked or open flame is, of course, the match and cigarette lighter.

The same cannot be said about heat sources other than flames, because, as the number of naked or open flames declines, the use of other heat-producing appliances seems to increase. This means that the higher temperatures necessary for a substance to reach its self-ignition temperature may occur by heat-transfer from hot appliances, friction heat, or chemical heat (self-heating), more often than was previously the case.

A further method of ignition utilizes fairly low temperature heat sources and yet does not need the application of a flame for combustion to occur. This phenomenon is known as pyrolysis and occurs when carbonaceous material is placed next to a heat source (steam pipe, etc.) for a prolonged period. The time period can be weeks or even months. If the material does not have an opportunity to reabsorb moisture regularly, because of the continuous heat or because the point of heating is too well insulated, then it can change its character to form pyrophoric carbon. This carbon will ignite at a lower temperature than the material it is derived from and may self-ignite.

In some cases, because of the insulation around it, which may limit the oxygen available, the material will not flame but will decompose due to combustion.

Some examples of sources of ignition, flame temperatures, burning characteristics of materials, etc., are given in the following tables.

TABLE 1.1 Examples of Sources of Ignition

ASHES	
SOOT	
CHIMNEY/STOVE	sparks from
PIPE/FLUE	overheating
INTENTIONAL BURNING	of building
SPREADING	of grassland
	of rubbish in the open
	of rubbish in the incinerator
NAKED or OPEN FLAME	matches
	candle
	taper
	solid fuel fire
GAS (all types flammable)	bunsen burner
	fire
	heater
	cooker
	welding/cutting
	soldering
CIGARETTES/PIPE/CIGAR	
OIL	lamps
	heaters
EXPLOSIVES/FIREWORKS	
ELECTRICITY	lead to apparatus
	lead/circuits in apparatus
	normal heat from apparatus
HOT METAL	
STEAM PIPES	
SELF HEATING	
SUN RAYS	
LIGHTNING	
STATIC ELECTRICITY	

TABLE 1.2 Examples of the Temperatures of Some Sources of Ignition

FLAME/SPARK SOURCES	°C	°F
Candles	640-940	1184-1724
Matches	870	1598
Manufactured Gas	900-1340	1652-2444
Propane	2000	3632
Light Bulb Element	2483	4501
Methane	3042	5507
Electrical Short Circuit or Arc	3870*	6998

NON-FLAME SOURCES	°C	°F
Steam Pipes at normal pressure	100	212
Steam Pipes at 10 lb/in^2	115	239
Light Bulb, normal	120	248
Steam Pipes at 15 lb/in^2	121	250
Steam Pipes at 30 lb/in^2	135	275
Steam Pipes at 50 lb/in^2	148	300
Steam Pipes at 75 lb/in^2	160	320
Steam Pipes at 100 lb/in^2	170	338
Steam Pipes at 150 lb/in^2	185	365
Steam Pipes at 200 lb/in^2	198	390
Steam Pipes at 300 lb/in^2	217	424
Clothes Iron	232	450
Steam Pipes at 500 lb/in^2	243	470
Steam Pipes at 1000 lb/in^2	285	545
Cigarette, normal	299	570
Soldering Iron	315-432	600-810
Cigarette, insulated	510	950
Light Bulb, insulated	515	960

*Copper wire burns through at 1095°C, 2003°F.

TABLE 1.3 Example of Pyrolysis of Timber

Prolonged exposure to a temperature of 80°C (176°F) to 110°C (230°F) can start the reaction	
At 150°C (302°F) change of state starts	
At 230°C (446°F) there is external browning	Self ignition possible
At 270°C (518°F) pyrophoric carbon is formed	
At 300°C (572°F) the wood becomes charcoal	

TABLE 1.4 Examples of Spontaneous Heating

Many substances when in contact with or mixed with certain other substances start to self-heat, i.e. the temperature of the combined substances rises. If, when this reaction starts, there is insufficient ventilation to allow cooling air to circulate, then the substances will ignite when the lowest self-ignition temperature of either of the two substances is reached. In certain cases, the reaction causes the heat build-up to occur so quickly that even good ventilation will not prevent ignition.

Vegetable oils soaked into waste rags are a good example of mixed substances which heat spontaneously to ignition and it is interesting to note that oils with iodine values above 100 are more susceptible to self-heating. The higher the iodine value, the greater the danger.

The following list of oils is in descending order, the more dangerous ones being shown first: Perilla, Linseed, Stillingia, Tung, Cic, Hemp-seed, Cod-liver, Poppy-seed, Soya-bean, Seal, Whale (sperm), Walrus, Maize, Olive, Cottonseed, Sesame, Rape, Castor,

It should be noted that many of the above oils are common in domestic occupancies as well as in industry (Linseed oil, for example, as well as being used for preserving leather, is a major ingredient of many polishes).

In addition to the aforementioned, there are, of course, several chemicals which create heat when mixed, sometimes with the violent release of hazardous gasses.

TABLE 1.5 Examples of Ignition Temperatures of Certain Materials

THE PARAFFIN SERIES

Material	Flash Pt.		Self-Ign.Temp.	
	°C	°F	°C	°F
Methane	Gas		536	999
Ethane	Gas		514	952
Propane	Gas		468	871
Butane	- 60	- 76	406	764
Pentane	- 40	- 40	309	590
Hexane	- 22	- 7	234	455
Heptane	- 4	25	223	435
Octane	13	56	220	428
Nonane	31	88	206	402

With this example it should be noted that as the flash point temperature increases the self-ignition temperature decreases for the same substance.

PLASTICS in FURNITURE (Reported in FP177, Feb 1977, F.P.A.)

Source of Ign.	Material	Results
Cigarette	Standard polyether medium density foam	Did not smoulder in usual way, but flamed after 70 min
Cigarette	Treated polyether foam	Self-extinguished
Cigarette	Rubber latex	Smouldered, then flamed
3kW electric fire within 20 cm	Standard polyether foam	Ignited, when foam reached 265°C (510°F)
3kW electric fire within 20 cm	Treated polyether foam	Did not ignite
3kW electric fire within 20 cm	Rubber latex	Ignited easier than standard polyether foam

TABLE 1.5 (continued)

The ignition and subsequent burning of furniture containing foamed plastics is dependent not only on the type of the foam but also on the type of covering used as the top surface.

The style of the furniture, whether a chair or a bed, and the location of the source of ignition will also affect the ignition

FLASH POINTS OF SOME ALCOHOLIC DRINKS

Whisky	28°C	(82°F)
Brandy	29°C	(84°F)
Gin	32°C	(90°F)
Sherry	54°C	(129°F)
Port	54°C	(129°F)

SELF-IGNITION TEMPERATURES OF TIMBER

TYPE	°C	°F
Various	180 to 350	356 to 662
All Timbers ignite after 30 sec exposure to	430	806
Wood-fibre board	215	420
Cane-fibre board	240	464
Melamine wood-flour filled	500	932
Melamine mineral filled	610	1131

TABLE 1.6 Examples of Temperatures at which Substances will Explode (without the application of a flame)

SOLIDS	°C		°F	
Gun Cotton (loose)	137	139	280	283
Cellulose Dynamite	169	230	338	446
Blasting Gelatine (with Camphor)	174		345	
Mercury Fulminate	175		348	
Gun Cotton (compressed)	186	201	368	395
Dynamite	197	200	388	392
Blasting Gelatine	203	209	397	410
Nitroglycerin	257		495	
Gunpowder	270	300	518	572
GASSES				
Propylene	497	511	925	950
Acetylene	500	515	932	960
Propane	545	548	1013	1017
Hydrogen	555		1031	
Ethylene	577	599	1070	1110
Ethane	605	622	1121	1150
Carbon Monoxide	636	814	1176	1495
Manufactured Gas	647	649	1196	1200
Methane	656	678	1212	1250

The higher temperatures are applicable when the heat rise is very rapid, i.e. if the rate of rise is slow then the explosion will occur at the lower temperatures.

It must be remembered that the figures quoted are the temperatures at which the substance itself explodes and not the temperature at which its container ruptures with the possible subsequent ignition of the contents.

TABLE 1.7 Examples of Burning Characteristics of Man-Made Materials

Trade Names	Chemical Form	Characteristics
Avisco, Bemberg	Rayon	Flammable (as cotton)
Dacron, Kodel, Fortrel, Vycron	Polyester	Moderately flammable melts, drips
Celanese, Chromspun	Acetate	Moderately flammable melts, drips
Orlon, Acrilan, Zefron, Creslan	Acrylic	Moderately flammable melts
Nylon	Polyamide	Slow burning melts, drips
	Polyethylene Polypropylene	Slow burning melts, drips
Dynel, Verel	Modacrylic (modified Acrylic)	Flame retardant melts
Saran, Ravana	Vinylidene Chloride	Flame retardant melts
Nomex	Polyamide	Flame retardant does not melt

Chapter Two *Arson: Reasons, Motives, Methods*

THERE are probably as many different reasons, motives and methods for fire raising, if looked at in detail, as there are people responsible for causing the fires. However, from the investigator's point of view, it is possible for all of these to be divided into six broad categories:

(1) To gain financially
(2) To conceal another crime
(3) To destroy/protest
(4) To become a hero
(5) To fulfil a need (mental disorder)
(6) Boredom

and these are amplified as follows.

1. Financial Gain

Under this category of fire raising, the arsonist usually tries to achieve four distinct aims:

(a) Firstly, the source of ignition is programmed to operate when he is remote from the premises. This ensures that any witnesses at the scene of the fire will not connect the outbreak with anyone seen leaving the premises; and it allows the arsonist to furnish an alibi for the time period when the fire occurred. Timing devices, some of which are discussed later, are often used to fulfil this aim. Of course, if the property to be destroyed is readily transportable—such as a motor vehicle, boat or caravan/trailer—the arsonist may remove it to a remote area where there are unlikely to be any witnesses. Once there, he can ignite a fire manually and remain at the scene to ensure complete destruction, thereby dispensing with the need for a timing device.

(b) Secondly, arrangements are made for the initial fire to spead rapidly and in some cases to spread in a particular direction. This may be done simply by positioning the initial fire where normal or forced ventilation, convection or conduction currents will cause spread, but more often several small fires are started at different locations or trails are used. Trails are formed in a variety of ways using many different materials too numerous to list. However, as a guide to the investigator it can be said that they fall into two basic sections:

(1) Those which use materials normally found on the premises, such as textiles in a draper's shop, flammable liquids from a leaking container in a motor repair garage, etc.; or

(2) Those which are manufactured as fire trails by the arsonist, such as rope soaked in a flammable liquid or streamers of toilet paper, etc.

(c) Thirdly, an attempt is made to delay discovery of the fire for as long as possible. The method used to achieve this aim will vary from building to building but it may take the form of one of the following methods or a combination of two or more of them:

(1) The fire is set in a building or part of a building not visited frequently by people in the premises, and not visible to members of the public from outside.

(2) The fire is screened from outside the building by closing curtains, stacking stock in front of windows, etc.

(3) Timing the outbreak so that it occurs during a time when there is little chance of it being discovered.

(4) Isolation of any automatic fire and intruder alarms and isolation of the manual/electrical fire alarm.

(5) Isolation of any fixed fire-fighting installation, such as automatic sprinklers, fixed carbon dioxide systems, etc.

(d) Finally, the fire will be set in such a way as to attempt to destroy all evidence of intentional ignition or so as to make the ignition appear accidental.

It will have been noticed, that in order to set a fire with any real chance of final success, i.e. gaining financially, the arsonist requires time, knowledge of the premises, knowledge of the processes carried on, and knowledge of the protection systems installed. Obviously, the person who has the most time and knowledge and generally the most to gain from subsequent insurance claims, sales of damaged stock, etc., is the owner or occupier. It is very important, therefore, for the investigator to look closely at all the physical evidence and to keep an open mind to any possible motive for arson. It may well be that the owner has not the technical knowledge to set a fire successfully, but the investigator should not overlook the possibility of an accomplice who might gain indirectly or someone who will arrange a fire in any premises for a fee. Alternatively, the owner, realising his limitations, might purposely make a very clumsy and amateurish attempt at fire raising and then stage a break-in to give the impression that the fire was caused by the supposed intruder.

It must always be remembered that arsonists are very ingenious, and even the most unlikely method of setting fire may be used, if it suits a particular purpose.

The following reasons for arson for gain are by no means exhaustive but are intended to stimulate thought in the investigator, if he finds it necessary to investigate the fire by motive:

Business

(1) Business poor and alternative to bankruptcy is arson.
(2) Cash flow problems resulting in large stock without buyers or difficulty in purchasing raw materials.
(3) Unable to meet fire or other safety requirements.
(4) Unable to fulfil existing contracts, particularly those containing penalty clauses for late delivery.
(5) Premises, machinery or commodity obsolete, with little chance of being able to be brought up to date.
(6) To put a competitor out of business or force him to lose an existing or possible order.

House

(1) Cannot dispose of property quickly enough to take advantage of job offer in another locality.
(2) Cannot dispose of property at right price because of condition of property or because neighbourhood has deteriorated environmentally.
(3) Due to be evicted through not paying mortgage.

Motor Vehicle, Boat, Caravan

(1) Unable to afford repairs.
(2) Special or foreign model depreciating too quickly.

(3) Running costs more than expected.
(4) Cannot keep up payments.
(5) Wants a new model.

2. Concealment of Crime

The use of fire in crime is becoming more prevalent; and in most cases the fire is started by the criminal while he is still on the premises. So, while the investigator will not find the remains of any timing devices, there will probably be ample evidence of trails or several points of origin or disruption of the contents of the room to form a bonfire, which all point to malicious ignition.

It seems that this type of arsonist is attempting the destruction of evidence of his personal involvement in the premises, obviously hoping that the fire will obliterate all fingerprints and cover up any signs of forced entry, rather than trying to make the fire look accidental.

A further type of fire for concealment is when an employee is responsible for falsifying records, account books, petty cash vouchers, etc., and, in fear of the discrepancies being discovered, sets a fire to destroy the incriminating evidence. An employee may also wish to destroy any adverse reports on his ability, etc., filed in the company personnel department, particularly if a new boss is to be appointed. In fires such as these, the prime suspects are the people with knowledge and access, and, as many companies now keep copy records, particularly if a computer is used, the investigator should study these for the names of people who may have had a motive for setting the fire.

In addition to being used to conceal a crime, fire is sometimes an active part of another crime, e.g. extortion and murder and these are dealt with separately below:

Extortion

In many fires, set as a warning or punishment by a person demanding money or other services off a property owner, the investigator will not be called in, because the victim, in an attempt to keep the incident as quiet as possible, will extinguish the fire without calling the fire service, or he will not make an insurance claim, or he will claim that he started the fire accidentally. On most occasions, fires started for extortion purposes will be relatively small (it is of little use to destroy a man's business, if the purpose is to obtain money from him). However, this cannot be taken as a definite rule particularly if the object is punishment; for example, one unit of a business with several outlets may be fired to destruction, thus allowing the criminal to continue to gain from the remaining outlets, or in an area of several individual small businesses, one may be destroyed as a warning to the others.

Murder and Suicide

Happily, there is little evidence to suggest that fire is used regularly as a weapon, although death is often the unintentional result of setting fires for other reasons, but it is used as a weapon too often for this motive to be ignored.

In most cases, the means of setting a fire for this purpose will be readily obvious as intentional ignition, since this type of arsonist tends to use fairly large quantities of flammable liquids, sometimes even resorting to throwing a "Molotov cocktail" type petrol bomb. If the object of the fire is to dispose of the body or bodies after a murder, then, unless the murderer really knows what he is about, there will always be sufficient tissue left for a pathologist to determine the cause of death or at least what is not the cause of death. It is not uncommon for the murderer to commit suicide after setting such a fire, sometimes involving his own body in the fire.

When suicide is the object of arson, it can take several different forms. Sometimes the arsonist/victim will take drugs or do some other act, before setting the fire, and sometimes he will simply light a fire. When fire is the sole weapon, it may be direct, with the person covering himself with a flammable liquid and then lighting it, or indirect when the person sets fire to furniture, etc., and waits for the smoke and toxic gases to render him unconscious.

At all incidents involving fatalities, the cause of death determined by a pathologist should always be sought by the investigator, before he finally states the cause of the fire. The reason for this is simply that either one can be responsible for the other. The fire, or fumes from it, may have been the cause of death, so the investigator must look for a cause of fire. Alternatively, if the deceased died from some other cause, he may, at the moment of death, have dislodged several articles and thereby brought some combustible material into contact with a source of ignition. A pan of cooking fat may have been switched on or a cigarette may have been left alight when death occurred and being unattended may have caused a fire.

3. To Destroy/Protest

A wide range of emotions is responsible for making people choose arson as a means of protesting or simply as a means of causing wanton destruction, and some of the reasons for those emotions are given below:

(1) Disagreement/ argument
 —person to person
 —organization to worker
 —organization to organization

(2)	Admonishment	—boss to worker —supervisor to subordinate —organization to member
(3)	Rejection	—person to person —organization to member —organization to prospective member
(4)	Termination of employment	
(5)	Civil disorder/ riots/terrorism	—religion to religion —sect to sect —organization to government

Fire makes the small and inadequate person very powerful, and it is always the most readily available weapon. As an example, a common incident of fire which can be attributed to any one of the first three circumstances in the destruction of school property and buildings by pupils.

From the investigator's point of view, these are easy fires to deal with, as there is not usually any attempt to disguise the fact that the fire was started deliberately. In fact, in many cases, it will be a momentary flash of anger on the part of the arsonist which, if the circumstances are right, will result in him firing the premises. In other cases, the firing is a result of the arsonist feeling an increasing resentment some time after the initial incident, particularly if his fellow workers or acquaintances keep referring to it.

During periods of civil disorder, simple but effective short-term timing devices are often used, not to disguise the fact that the fire was intentional, but to allow the terrorist time to escape from the immediate area. Often, the terrorist group will claim responsibility for the outbreak, and may even give warning that an incendiary device has been planted. Such devices can be as small as a cigarette packet or as large as a suitcase.

4. The Hero

As a purely personal observation not backed up by any statistics or psychiatric knowledge, it appears that this "hero syndrome" occurs most frequently when the community or environment *related to the Hero* is of limited scope, quiet and uninteresting. Thus even in a large city, if the individual's circle of involvement is limited, he may develop the "hero syndrome".

The Hero perhaps feels stifled or wishes to "prove" himself to a girl or his boss or the community as a whole, and if very little happens within his limited sphere to satisfy those wishes, then he decides to create a situation which will allow him to become a Hero. Often, this results in the Hero starting a fire and then extinguishing it or raising the alarm or performing a rescue.

Usually, these fires are small, because (of course) the Hero wants to discover them before anyone else. The danger is, however, that having been a Hero once, the arsonist tends to repeat the performance, and will in most cases continue to do so until he is arrested or he starts a fatal fire and the shock cures him.

The idea of using fire in this way might follow a fire caused by "normal" circumstances which resulted in a bravery award, wide publicity or excessive praise for the person discovering the blaze, particularly if the Hero was nearby or actively involved.

Some signs of this motive for arson: (1) several small fires in an area which are discovered before they have a chance to develop; (2) unusual attitude of the person discovering the fire, fully dressed at night, projecting himself as Hero, etc.; (3) same person at each of a series of small fires.

It is possible, of course, that the fire will develop more quickly than the Hero intended, even resulting in the total loss of a building or in a fatality. When this happens the Hero may become fearful or remorseful and close questioning will often lead to a confession.

5. Fulfil a Need

In general, this section covers those people who suffer from some problem, either mental or sexual, which is expressed or released when the person sees flames from a fire. The investigator should always check into the background history of employees or any other person who seems to be unusually excited while at the scene of a fire, particularly if his excitement wanes when the flames die down or are obscured by smoke. People who loiter near fire stations or follow fire applicances to fires should also be treated with suspicion.

It is also possible to place children playing with fire under this category, as there is usually little reason for a child to light a fire (with the exception of pupils setting fire to school property as mentioned earlier) other than to watch the flames or see a piece of paper transformed into a pile of ashes.

6. Boredom

The signs and systems of firing for this reason are similar to those described in Section 4, *The Hero*, above, and in many cases are caused by a similar person to the Hero. It is most common where people are doing a job which is either repetitive or does not have sufficient mental or physical stimulus to satisfy the person involved. Some people will cause mechanical damage, such as stopping the conveyor belt in a factory production line; other people start fires.

Chapter Three *Arson: Timing Devices*

THE METHODS outlined below are a sample of the type of timing device which may be used by arsonists, particularly those who come within the first category discussed in the previous section, i.e. to gain financially. The advantages and disadvantages discussed are from the point of view of the arsonist, not of the investigator.

1. Remote Control

(a) Telephone

A telephone in the area in which the fire is to start can be rigged so that when a call is made, the bell and striker close an electrical contact to cause a spark or to bring a heating circuit into use, depending upon the method of ignition necessary for the material to be ignited. As an alternative, a friction device and match may be utilized.

In some cases, an extra alarm bell or other type of sounder is fitted to the telephone installation, and if this is used instead of the receiving instrument, the resulting fire could be remote from any locations normally associated with telephones.

A further variation is where a private exchange can be switched to provide for different extensions to ring if a call is received when the exchange is unmanned. This allows the arsonist a much wider scope in choosing the best area in which to set the fire, or he can rig several extensions to start a number of fires simultaneously.

One advantage of this method of firing is that the arsonist can choose the exact time at which he wants the device to operate and he can be miles away when he dials the number (he could even be overseas). Of course, there is always the chance that someone else might call before the planned time, thereby nullifying any alibi; and it might prove difficult to ensure complete destruction of physical evidence at the point of ignition.

(b) Radio

Mainly due to the popularity of radio-controlled model boats, aircraft and motorcars, there is a variety of very compact, highly sophisticated control systems which can be readily adapted to operate an incendiary device at a given point. There are, however, three principal disadvantages with this method; (1) the screening of the structure might prevent the system operating; (2) the arsonist has to remain relatively close to the premises due to the limited range of the systems readily available; and (3) in any premises other than a model shop or a private household, discovery of such a system would of course immediately arouse suspicion.

2. Short-Term Timing

(a) Fuses

(1) *Candles.* Probably the most basic and readily available timing device is a candle with its base set in combustible material. When it burns down, the flame reaches the material and causes ignition.

Candles of different cross-sectional area or made from different waxes burn at different rates giving a wide time-range; and as those of similar manufacture and cross-sectional area burn at approximately the same rate, the time can be varied simply by altering the length of the candle.

Disadvantages of candles are that the flames may be extinguished before reaching the combustible material, therefore leaving evidence of a plant; or the candle may fall over and ignite the material prematurely. Also, even after a successful fire, chemical analysis of the area around the point of ignition can show traces of wax and therefore give rise to suspicion.

(2) *Cigarettes.* A cigarette used as a fuse is not as foolproof as a candle, since it is at a much lower temperature and does not, under normal circumstances, flame. For it to be utilized with any certainty, it is usually laid in a box or book of matches, causing ignition when the cigarette has burned away sufficiently for the hot tip to come into contact with the heads of the matches.

The disadvantages are, again, that the cigarette may be extinguished, and that the flaming caused by the matches, while generally hotter than a candle, is relatively short lived and must, to be successful, ignite the material quickly.

Cigarettes, when insulated by furniture upholstery or bed clothes, often build up sufficient heat to cause ignition.

However, this method is not used by arsonists, as it cannot be relied upon to cause ignition.

(3) *Manufactured Fuses.* This type of fuse is not often used by arsonists, as there is a limited market for its commercial and industrial use and it is not therefore readily available. In addition, its use can easily be detected by chemical analysis.

(b) Liquids

(1) *Hazardous Liquids.* The most common type of liquid timing involves the use of two liquids, one of them corrosive, which when mixed cause an exothermic reaction. These liquids, are separated by an easily corroded membrane. By altering the thickness or material of the membrane, the length of time required for the corrosive liquid to eat through it and come into contact with the other liquid is correspondingly altered.

(2) *Safe Liquids.* A second method involves the use of those hazardous substances which react violently when exposed to the air and which are normally stored under a liquid. Using this method, the substance is placed in a combustible container and a hole is made to allow the storing liquid to drain at a pre-arranged rate. As soon as the substance is dry it starts to react and will, on ignition, set fire to the container.

Of course, this drain system can also be used as a variable weight on a see-saw or balance to raise or lower an arm and thereby operate a switch, an ignition device or pour a secondary liquid.

(3) *Water Reactive Substances.* Finally, there are those substances which react violently with water. Contact can be made by raising the level of water in a container in which the substance is raised on a central tower. Careful research must be carried out before setting this type of ignition device, as

the water which starts the fire can, under certain circumstances, also extinguish it. In the case of a boat, it is a simple matter to introduce a small leak in the hull or not to pump out normal leakage which seeps into the bilges.

(c) Spontaneous Ignition

Many oils, particularly those with high iodine values, are liable to ignite spontaneously when soaked into carbonaceous or cellulosic material and placed in poorly ventilated areas. For example, linseed oil (which is probably more susceptible to self-heating than most oils) soaked into cotton waste and placed in a combustible container, such as a shoe box, will create sufficient heat to ignite the cotton waste and in turn the shoe box.

Varying the quantities of oil and cotton waste can vary the time taken for ignition.

A similar effect can be obtained by using an oxidizing agent in conjunction with combustible solids or liquids.

(d) Electrical Installation

Use of the normal electrical installation in the premises has the advantage of being a "natural" cause, and, therefore, might not attract undue attention from the investigator. However, it is not certain that the fire will start on cue or even start at all, due to fluctuations in the supply, power cuts, operation of fuses, etc. There is also the possibility that an insurance company might not pay out, if there are obvious signs of "negligence" in the care and maintenance of the installation.

With the electrical installation, it is possible to cause overheating by simply overloading the system, but to overload a system it is necessary to fit a higher rated fuse than normal or to bypass the fuse completely. Whichever method is used, there is always evidence left for the careful investigator.

(e) Clocks

(1) *Alarm Clocks.* Most alarm clocks can be rigged in the same way as the telephone mentioned above, i.e. by the sounder closing an electrical circuit or operating a friction device. An advantage over the telephone is that the device can be set anywhere in the premises. However, against that, is that the remains of an alarm clock found in an area where one would not usually be used would immediately arouse suspicion.

(2) *All Clocks.* Utilising an electrical circuit, the face and hands of any traditional clock can be wired so that they become a switch. A location on the face is chosen as one contact point and one of the hands for the other (the hand not being used is usually removed). When the hand reaches the point on the face, the circuit is closed creating a spark or switching on a secondary device. The switch can be made to operate and then switch off as the hand moves on or a stop can be made on the face to ensure that the switch remains closed.

(3) *Switch Clocks.* Of course, in some situations clocks are used normally to control the times at which machinery, heating systems, cooking ranges or display lighting operate. By inducing a fault into the equipment to be operated and then by-passing any safety cutouts the operation of the time switch can cause ignition or at least start the heating necessary to cause ignition.

(f) Heat Sources

(1) *Radiation.* One method of timing which, when investigated, could be determined as non-intentional, is the ignition of material by radiated heat, e.g. wet clothing drying in front of a fire, furniture too close to a fire, or an electric fire placed too close to furniture or any other combustible materials. A few simple experiments with the heat source and the material to be ignited will enable a time scale to be prepared related to the degree of dampness of the material and its distance from the heat source. When this type of ignition is suspected, the investigator must examine carefully any possible motives for arson.

(2) *Conduction/Convection.* A more direct method can be utilised in any premises where combustible materials are heated in the usual production/preparation processes, e.g. fat-frying ranges, glue manufacture, paint-drying ovens, etc. The method is simply to cut out any thermostats and safety timing mechanisms and leave the equipment in its heating mode, resulting in overheating and ignition of the substance being treated.

(3) *Direct Burning.* Where flammable gasses or liquids are present in any premises, a small fire or some other source of ignition can be set and the valves of the gas cylinder or liquid container turned on. The intention is that the gas or liquid will ignite on reaching the set fire. Two disadvantages of using gas are that it always burns back to source and, if the explosive range is reached before the gas ignites, it is possible for an explosion to occur without sustaining combustion.

3. Long-Term Timing

The following methods are not commonly used because they do not have the same degree of probability as others mentioned and they can usually only be used when climatic conditions over a long period can be forecast with certainty.

(a) *Sun's Rays.* If direct sunlight through a window is a regular feature of a particular season, a lens or concave mirror can be positioned so as to focus the sun's rays for the longest period possible. After the focal length of the lens or mirror is determined, it is a simple matter to select some combustible material most readily ignited by this means and place it within the focal length. This method is particularly suitable where the premises are to be unoccupied for a long period, with all normal services/utilities cut off, e.g. during a summer break.

(b) *Rain.* As mentioned under 2(b) (3) above, there are substances which react violently with water, and a variation on the leaking boat theme is a leaking roof or faulty gutter, gully or downspout. The substance is placed in position, the premises shut, and the arsonist simply waits for rain.

Chapter Four *Explosions*

FOR FIRE investigation purposes, explosions can be divided into two basic categories: a fire explosion (where the substance which explodes is combustible), and a non-fire explosion (which is usually the result of an increase in pressure within a vessel, above that which it is designed to withstand).

Both of these types of explosions have certain characteristics in common. They will always follow the path of least resistance, even if the difference in resistance is minimal, and the power of the two explosions is always greatest at the point of origin. This initial power diminishes in proportion to the distance from the point of origin and in relation to the size and number of obstacles in the path of the blast.

Accordingly, with both fire and non-fire explosions, the effects from them will be worse near the point of origin and the blast will travel for considerable distances along corridors, ducts, shafts and open areas. The blast will, in many cases, bypass plant, side rooms with closed doors, etc., because of the difference in resistance.

In addition, as the distance from the point of origin of a fire explosion increases, only the more flammable material will ignite and not material which requires a large amount of heat energy.

1. Non-Fire Explosion

The investigator is interested in the non-fire explosion for two reasons: firstly, when the explosion occurs before the fire, and flying debris causes a spark or a short circuit in electrical apparatus or anything else sufficient to act as a source of ignition; and secondly, when the blast or debris causes damage to other plant, which either allows a hazardous substance to escape into an environment where it can be ignited or simply releases an already-burning process material.

2. Fire Explosion

When the explosion is a fire explosion, the material which explodes is inherently combustible and explosive, and does not explode solely because of an increase in pressure in its container. Fire explosions can, for investigation purposes, be divided into two further categories, viz. one which results in a sustained fire, and one which results in a flash fire.

The type of fire which results is dependent entirely on the combustibility of the surrounding material in relation to the heat energy given out by the explosion. For example, if a fire explosion of a particular magnitude will produce sufficient heat energy to ignite certain made-up textiles but not baulks of timber, then a sustained fire will result if those textiles are within range of the effects of the explosion. However, if instead of those textiles, only baulks of timber are within range, then the result will be a flash fire.

Whichever of the two sub-divisions is the result, it will leave its own distinctive sign for the investigator to interpret.

(a) Flash Fire

(1) Only more combustible material affected.
(2) Signs of fire damage along paths of least resistance (rapid temperature rise, short period of burning).
(3) Damage inconsistent with a sustained fire—e.g. after a gas explosion in a domestic kitchen the plastic case of a radio and a baby's waterproof pants had melted and the person in the room suffered slight burns to the face. However, although a large window was blasted out of its frame, the lace curtain across the window was not damaged by either heat or the blast.

(b) Sustained Fire

(1) Possibly several seats of fire formed.
(2) Point of explosion may show a large fire-damaged area with signs of a rapid temperature rise.
(3) Any additional seats of fire formed will show smaller areas of fire damage with signs of a rapid rise.
(4) The fires may spread rapidly to join one another—e.g. a leakage of flammable gas into a small timber cupboard can, on ignition, cause the same depth of charring (assuming same timber) on all sides, top and bottom of the cupboard.

3. Time of Explosion

It is important to ascertain as far as possible the time that the explosion occurred. If a fire is discovered after an explosion has happened and no fire was visible before, then it is reasonable to assume that the explosion contributed in some way to the start of the fire.

However, while it is definite that any explosion occurring during a fire could not have started the blaze, but could only intensify the fires, the material which exploded could have been involved at the start of the fire. For instance, if the contents of a gas cylinder leak through a hole in the structure of the cylinder or through a valve, and the leakage is ignited, then the gas becomes the material first ignited. If, after this initial fire has started, no compensatory action is taken, the cylinder will heat and could explode. This explosion will have occurred as a result of the fire, but if the original fire was not discovered before the explosion and subsequent spread of fire took place, any witnesses would probably assume that the explosion caused the fire instead of the opposite being the case.

4. Flash-over

A flash-over can occur in any type of fire and can, to the inexperienced, appear to be an explosion. Accordingly, it is important that evidence obtained from witnesses is precise as to whether the rapid propagation of flame was from an explosion or a flash-over. As a very rough guide, if there was a large volume of smoke present before the "explosion", then it may have been a flash-over: alternatively, if there was a lot of flame visible before the "explosion" then it is doubtful if it was a flash-over.

Briefly, "flash-over" is a term used to describe what happens when the temperatures of the contents of a room or building are raised to the ignition temperatures, but do not ignite because of an oxygen deficiency. When the room or building is ventilated, the inrush of air allows all of the contents to ignite simultaneously, sometimes with explosive force.

5. Examples of Explosions

(a) Dust Explosions

Dust explosions are caused when a combustible dust is suspended in finely divided particles in air and a source of ignition is introduced. These explosions most commonly occur in cyclones and extraction ducts, where it is relatively easy for the fast-moving machinery to overheat or for the accidental introduction of a foreign body to cause a spark.

Another cause of explosions in dusts is when a smouldering fire in a dust pile is either disturbed or creates its own minor explosion, thereby releasing a dust cloud which the smouldering fire ignites.

(b) Gas Explosions

The majority of gas explosions involve manufactured gas, natural gas or liquified petroleum gas. However, most leakages of gas are either ignited soon after the leak occurs and burn as a "blow torch" or are discovered and shut off before ignition takes place.

For an explosion to occur the gas must be mixed with air in such proportions by volume to be within the upper and lower explosive limits of the particular gas. Once within these limits, the introduction of a spark or flame will cause an instantaneous explosion. Below the lower limit, there is insufficient gas in the mixture, and above the upper limit there is insufficient air in the mixture for combustion, and therefore explosion, to take place.

(c) Liquid Explosions

One type of explosion which is very powerful but quite rare, involves liquid, the vapour from the liquid, and a pressure vessel. It is known as a bleve (boiling liquid expanding vapour explosion).

This type of explosion occurs when the pressure vessel containing the liquid is heated from an external source. As the temperature of the tank and its contents rises, so too does the internal pressure causing the tank to rupture, usually above the liquid level.

When the break occurs, the pressure within the container suddenly drops, thereby allowing the liquid to boil, vaporize, and expand with tremendous force. The fire which caused the heating of the container provides the source of ignition for the gas cloud, resulting in a fireball.

This series of events can happen even with a large capacity storage tank fitted with relief valves. If the heat input produces more vapour than the pressure relief valves can cope with, then the container will eventually explode. It must also be remembered that the pressure needed to burst a metal container is considerably reduced by the weakening effect of heat on metal.

Chapter Five *Limits of Fire: Time Burning; Temperature Reached*

1. Rate of Growth

THERE are many variables which have to be taken into account when trying to determine the rate of growth of a fire, because, as the following examples show, a fire does not have a uniform rate of rise and the signs remaining can be confusing;

(a) A small fire may smoulder undetected for a considerable time and then suddenly, because of a pane of glass breaking or timber panelling burning through, it may become an inferno as it is fed an unlimited supply of air.

(b) A fire in a small room will heat its surroundings quicker than the same fire in a large room, because there will be a smaller quantity of material to absorb the heat and it will, of course, be closer to the heat from the fire.

(c) Windows will be protected from the effects of fire initially, if screened by velvet or woollen curtains. Conversely, if the curtaining is of a highly combustible nature, the windows may fail earlier than normal

However, in spite of these limitations and variations, the table opposite gives several indications of how the investigator can obtain a very rough estimate of the rate of growth of a fire.

2. Period of Burning

In the past, one method of determining the length of time a fire had been burning was to measure the depth of charring of timber. The depth in inches was then divided by one-fortieth (timber was supposed to burn at one-fortieth of an inch per minute) and the answer was taken as the length of time in minutes that the fire had been burning.

However, a new British Standard (BS 112 Part 4) gives the following burning rates for varying densities of timber:

(1) Density less than 420 kg/m^3 (26 lb/ft^3): 25 mm (0.98 in) in 30 min
(2) Density of 420 kg/m^3 (26 lb/ft^3) or more: 20 mm (0.79 in) in 30 min
(3) Hardwoods: 15 mm (0.59 in) in 30 min

The Standard also states that rapid fires and very high temperatures will increase the burning rates above the figures shown.

3. Temperature Reached

It is virtually impossible to measure the exact temperature reached by a free-burning fire because of the many variations

Material	Indications	Rate of Rise
GLASS	Clean breaks, possibly following lines of frame	Rapid (1-5 min)
	Crazed, with little or no staining from smoke	Moderate, with intense heat and little smoke
	Softened glass still held in position	Slow (prolonged), intense heat
	Heavy staining from smoke, no crazing	Slow, with heavy smoke
	(The investigator should bear in mind that some glass is tinted during manufacture to give a smoked appearance and he should, therefore, always examine any glass in neutral lighting before reaching any conclusions.)	
TIMBER	Sharp demarcation between charred area and clean timber	Rapid
	Hazy demarcation between charred area and clean timber	Slow
PLASTER	Off walls	Rapid
	On walls	Slow
	Mainly on walls, but off at, or above, point of origin	Moderate
	(These indications are obviously affected by the type of plaster, its thickness, its age and the quality of workmanship when construction took place.)	
WALL-PAPER	Black coating left on the wall, possibly with the pattern still visible.	Slow, with intense heat
	(What has happened is that the paper has been heated to decomposition without being touched by flame. This is know as pyrolysis: see Chapter 1, *Sources of Ignition.)*	

of fuel which may be involved and the construction and layout of the building in which the fire occurs. The investigator should be very careful, therefore, not to make any statements about the temperature reached which he cannot substantiate. For example, the inexperienced investigator, knowing that steel buckles and fails at about 550°C (1022°F) might assume that whenever buckled steelwork is observed at a fire the temperature attained was at least 550°C (1022°F). However, if one looks at the characteristics of heated steel a little closer, it becomes obvious that this may not be the case:

(1) The temperature may have reached a considerably higher level than 550°C (1022°F), but, because it was not sustained for a long enough period, there is no visible effect on the steel.

(2) If the ends of the steel are prevented from expanding laterally, the steel will buckle at a lower temperature than 550°C (1022°F) to cater for any slight expansion.

(3) If expansion areas are built in and the load on the steel is proportionally minimal, it is possible for it to expand and later contract without visible distortion.

(4) The durability of steel in fire is related to its thermal capacity and surface area, so that for the same temperature for a similar length of time, different sizes of steel will display varying degrees of effect.

Accordingly, if the substances listed on the next page show signs of having melted, the temperature given must be treated as only a rough guide and not as a definite value for the maximum temperature reached by the fire.

MELTING POINTS

	°C	°F		°C	°F
Expanded Plastics	121	250	Brass	871	1599
Tin	233	450	Silver	955	1751
Lead	330	626	Gold	1065	1949
Zinc	420	788	Copper	1082	1979
Aluminium	649	1200	Cast Iron	1232	2249
Glass	760	1400	Steel	1426	2598
Bronze	788	1450	Nickel	1482	2699

Probably, a more accurate figure can be obtained by examining any concrete which has been exposed to the heat. Tests conducted by the Fire Research Station in England showed that colour changes occurred in the concrete at certain temperatures as follows:

Colour	°C	°F
Pink, Red or Red/Brown	300	572
Grey (coarse aggregate may remain red)	650	1202
	(marked loss of strength)	
Buff	1000	1832
Sintering of concrete and mortar Fusion of bricks	1250	2282

After the heat part of the test was completed a cross-sectional cut was made through the test piece and it was observed that the above colours had "run" into the concrete similar to the way in which tempering colours run along steel.

The depths of the colours were:

Pink	Grey	Buff
75mm (3 in)	25mm (1 in)	6mm (¼ in)

In view of the above, it seems that even though the concrete may be severely smoke-blackened, it might be possible to clean the surface sufficiently to see a colour and thereby obtain a value for the temperature reached.

Chapter Six *Fatality*

WHEN a fatality occurs in a fire, it often makes the job of the investigator somewhat easier, as it is possible for a pathologist to obtain a substantial amount of information from an examination of the body.

It must be remembered, however, that in some cases there will not be any information available before the investigation starts regarding the possibility of a body being present. If the premises have been secured for the night, or a roll-call following evacuation suggests that all persons are accounted for, the firefighting crews may not institute a full-scale search and a body could remain in the fire area until long after the fire has been extinguished. In fact, tests have shown that the body fats can be rendered down and burned to destruction, particularly if the body is suspended above the flame by being held on the metal framework of a chair or bed; and if this has happened it may be difficult to discover the remains of a body.

Accordingly, great care must be taken when sifting through debris, particularly when carrying out the excavation and reconstruction technique described later. Immediately anything is discovered which suggests that there may have been a casualty, the surrounding area must be subjected to a meticulous search. In cases of multiple fatalities, it is imperative that all of the remains of bodies are accounted for, so that an accurate number of casualties can be determined. Sometimes the only way to be reasonably sure of the numbers involved is to count the remains of heads or skulls found.

If the body is discovered shortly after firefighting starts and there is no doubt that life is extinct, then it should be left where it was found until the investigation can be started. Of course, if there is any danger that the fire may spread to involve the body, then it should be removed, but its location, position of the limbs, and whether it was trapped by furniture, machinery or fallen debris, should be noted.

It is important to be thorough when a body is found, as in many cases the collection of information about the body and its location and past history will assist in determining the cause of the fire.

When carrying out the investigation, particular attention should be paid to the following points;

(a) The identity of the body. The investigator must not jump to conclusions. If he is informed that a guard dog (or some other animal) was in the premises and a torso is found, he must have the body checked to make sure that it is that of an animal and not the body of a child. Even if the body is that of an animal, the cause of death should be determined and an examination made to ascertain whether any drugs had been administred before the fire. Once the identity of the body is known, the same tests to establish cause of death and whether or not any drugs had been taken should be arranged.
(b) The position of the body in relation to the layout of the room and floor in which it was found.
(c) The position of the doors and windows, furniture or machinery in the room where the body was found.

(d) If the doors were open, closed or locked. If locked, what type of lock was used, and where was the key?
(e) If the fire was between the body and the exit which would be used in normal circumstances.
(f) If the windows were open, closed, locked, fitted with toughened glass, of sufficient size for the casualty to have passed through, or barred.
(g) If the body was trapped by furniture, machinery, fallen debris, etc.
(h) If any suicide notes have been left.
(i) If the casualty was mentally ill, physically disabled, under the influence of drugs or alcohol, currently under the care of a doctor.
(j) If the casualty was asleep.
(k) If the casualty was a child, whether left alone on this or any previous occasion; and if the child has a history of lighting fires.
(l) If anyone else involved in the premises has a history of lighting fires.

Finally, an account should be taken of any rescue attempts and the reasons for their failure.

When the above information has been obtained, the investigator should find out the recent history of the casualty to determine his trade or profession, his home and place of employment, and the time, date and place he was seen last. This will enable the investigator to ascertain if the casualty had a right to be in the premises or in the area in which he was found when the fire occurred, the time limits during which he attended the premises, and whether or not he was likely to have had anything to do with the fire occurring.

Chapter Seven *Photographs and Drawings*

WHEN an investigation has been concluded, the investigator needs to do two things: he needs to report his conclusions to the person or persons who required the investigation; and he needs to keep a comprehensive record of the investigation for future reference (it can take months or even years after the end of the investigation before a case is concluded in the courts or an insurance claim finalized).

For both of these requirements, he will almost certainly prepare a written report and may attach appendices relating to particular aspects of the case, the contents of the premises, or the pathologist's report. However, no matter how competent a writer the investigator may be, he will find it difficult, if not impossible, to convey accurately the description of the floor layout of a building, the production flow line in a factory or the position of a dead body in relation to other objects in the room, without supplementing his report with photographs or drawings or both.

1. Photographs

Photographs are particularly useful, as they can, in most cases, capture exactly the remains of the contents of the fire area, the extent of the fire, water damage, and the physical signs of whether doors were open or shut. Also, if the results of the investigation are to be presented in evidence, it must be remembered that some people cannot read plans or other drawings, but they can relate a report to photographs.

Accordingly, photographs should be taken as follows:

(1) External

(a) The external faces of the building(s) involved and any adjoining property, vehicles, etc., which could have assisted in the start of the fire or obscured it from a main thoroughfare or to which the fire could have spread.
(b) Any signs of break-in through fences or gates or into the building itself.
(c) Any external fire spread on the building first involved.
(d) Any signs of an external fire (rubbish burning, grass, etc.).
(e) Any footprints or tyre prints which give cause for suspicion.

(2) Internal

(a) General view of the remains in the fire area.
(b) Each stage of excavation (showing contents and layout of each level if possible).
(c) The area of origin, the point of origin, and first material ignited.
(d) Any trails.

(e) The condition of the electrical installation, gas, and any other services which might have had a bearing on the fire.
(f) Any abnormalities in the fixed fire-protection arrangements.
(g) Any malpractices in storage of hazardous substances, etc.
(h) Detailed picture of any bodies found.

2. Drawings

Drawings are useful in themselves and also as an aid for photographs. The plans drawn after a fire can be compared with previous plans of the premises to check on any alterations which may have been made, etc. The positions from which the photographs were taken can also be indicated on the plan.

The following items may be described in drawings:

(1) Structure

(a) Plans and elevations.
(b) Sections and exploded 3-D views.
(c) Details of internal linings such as suspended ceilings, false walls, etc.
(d) Wiring diagrams, pipe systems, etc.

(2) Plant/Contents

(a) Layout of plant and contents.
(b) Elevations of plant.
(c) Exploded views through machinery.
(d) Production flow line diagrams.
(e) Position of contents, doors, etc., in relation to the position of a casualty.

When using photographs and drawings it should be borne in mind that colour photographs give a more real picture of a scene, even though, with the effects of smoke, the predominate colours may be black and grey. It should also be remembered that black and white drawings done on tracing paper in ink are the easiest to reproduce and give a good copy.

It is essential that the investigator completes his recording by photographs and drawings as soon as possible, as subsequent cleaning operations, construction repairs, or demolition will remove the evidence very quickly.

It may be of benefit in certain cases, particularly when the investigator is called to the scene late, to obtain photographs from other sources such as the local newspapers, freelance and insurance photographers. In some cases, a video-tape recording or a cine film may be available.

Chapter Eight *Samples*

IT IS a good policy to collect samples from a fire scene whenever possible and, as with photographs, if the samples are not collected at the earliest opportunity, they will be lost forever. The investigator should, therefore, obtain as many samples as he considers necessary to assist him in determining the cause of the fire or to support any suspicion of a motive for arson. Remember, it is better to collect too much rather than not enough, excess samples can always be disposed of later.

1. Method of Collection

Great care should be taken on the initial approach to the fire scene, so that any evidence remote from the point of ignition is not damaged any more than is necessary to ensure extinguishment of the fire. As samples are obtained, the point where they were found should be indicated on a plan, so that they can in the future be related to the point of origin. Large items, such as machinery parts and other non-absorbent items should be marked with a tie-on label; other samples, such as pieces of timber, textiles, soil, etc., should be collected in a glass screw-top jar or a plastic bag wired shut. If it is possible and desirable to collect documents, this can be done by sandwiching the paper between two sheets of flexible transparent plastic.

All samples must be conspicuously labelled when they are obtained at the scene; and the label should show the place where found, the nature of the sample with its trade name if possible, and the fire identification (date and time of call, or fire number, or address of premises). As the samples will often be obtained in wet and dirty conditions, it is a good idea for the label to be made of white plastic and any information to be entered on it in chinagraph pencil.

2. Sending for Test

No matter how good a chemist the investigator may be, he should always arrange to have all chemical tests of samples carried out by a recognised testing or research laboratory, so that the chemist becomes a witness to prove or disprove any theories about the fire. Similarly, if the investigator requires any fire testing of samples to be undertaken, then he should send them to a fire testing laboratory which can scale down the area and quantities, simulate the layout and the method in which the goods were displayed or stored and introduce varying degrees of ventilation, heat and humidity.

Only a part of each sample taken should be sent for test or experimentation, then, if the sample is lost or destroyed, a second sample is available for test. Also, it may be that if there is any doubt about the validity of the first test, this can be dispelled by having a further part of the sample tested at another laboratory. In addition, the parts of the samples sent for test should be in similar containers to those mentioned above, but should be marked only with a code number relating to the investigator's report, and should not contain any information about the date, time or location of the fire.

3. Reasons for Collection

(1) On Site

(a) To test for the presence of accelerants and other substances which would not normally be present.
(b) To test for self-heating. (Where self-heating is suspected, a sample of the material at the point of origin and a sample from the outer edge of the material should be obtained for test and comparison.)
(c) To carry out experiments as to combustibility, flammability, etc.
(d) To identify the type of contents of each floor level, particularly important when a collapse has occurred.
(e) To assist in identifying suspects (threads of clothing, strands of hair, traces of blood near point of entry).
(f) To test for ease of salvage and reclamation (excessive salvage claims by the occupier could point to arson).

(2) On Body or Suspect

(a) Clothing for traces of fire, accelerants, etc.
(b) To compare mud and soil on clothing with that at the scene of fire.
(c) To find out identity of body.

Chapter Nine *The Investigation: Physical Evidence*

PHYSICAL evidence is that which is found at the scene of a fire after it has been extinguished. It is important for the investigator to make a very close study of such evidence in all cases, so that any deductions and statements that he might make in the future will be able to be substantiated by reasoned practical arguments. Sometimes, the evidence collected, and the deductions made, will not prove what the cause of the fire was, but will prove what could not have been the cause. All physical evidence must be considered against witnesses' statements and other documental evidence, before a final statement of cause is made.

The investigator should always work to a predetermined plan, to ensure that he covers all aspects thoroughly, yet does not go over the same ground more often that is necessary. One method which can be used is firstly to carry out a general examination of the premises as a whole and then concentrate on a detailed examination of the area of origin. The scope of both of these examinations is amplified below.

It will be noted that for both building fires and motor vehicle fires the "General Examination" starts by checking for arson. However, as stated in the Introduction, arson should not be given undue prominence. The signs which may indicate a possible arson fire are mentioned first, simply because they are the first items which the investigator can check as he is approaching the fire scene. If he does not carry out these routine checks at the earliest opportunity, the evidence may accidentally be destroyed, particularly as it could be remote from the heart of the incident.

1. Buildings

(1) General Examination

(a) Signs of Forced Entry

If there are signs of forced entry then it is a reasonable assumption that malicious ignition is a probable cause of the fire. However, the investigator must make sure that any signs of forced entry or breakages were not made by the firemen, policemen or other persons after the fire had been discovered or, indeed, by the fire itself.

Signs to look for:

(1) Broken doors, hatches, roof lights, pavement lights, windows, etc., or split or scratched frames of doors and other openings caused by a crowbar or some other type of lever.

(2) Broken glass: check if pieces of glass are on the inside or outside of the frame; check the edges of the pieces left in the frame to see if they are discoloured (depending upon the smoke density, heavily stained edges may indicate that the glass was broken before the fire started). The edges of broken glass inside the premises, from bottles and other containers, as well as internal glazing, should also be checked for discolouration.

(3) Footprints outside or inside or dried mud, etc., which has obviously been brought in from outside. Impressions on a recently vacuumed carpet or a silhouette made in spilt powder or on a damp carpet.

(4) Fingerprints are generally difficult to obtain after a fire, because of the action of the fire and the number of people involved in damage control and salvage work. However, if the fire has been of a minor nature, it is sometimes useful for the investigator to arrange for fingerprints to be taken by the police for checking against their records. Even prints of employees who have a right to handle certain material, may be useful, if it can be proved that their prints were left on a filing cabinet, safe door handle or office door handle after cleaners had dusted and polished at the end of the day. Prints of operators found on machinery after the maintenance department have completed a service are also suspect, as is evidence of an employee being in an area not normally visited by him.

(b) Aids for Fire Growth

The following signs may indicate arson or in some cases just negligence and as with the items mentioned above, the investigator must check that the results were not caused during the fire by firemen, etc. These signs are of obvious interest to an investigator, if they are part of an arson plan; but even if they are due simply to negligence, they may be of interest to the insurance company, who could restrict the amount of money paid out on the grounds that the owner was partially responsible for the extent of the fire.

Signs to look for:

(1) Fire resisting self-closing doors open (check to see if the self-closing device has been tampered with to make it inoperative).

(2) Fire separation doors prevented from closing by stock or wedges (check fusible link or other method of release to see if wired together or painted to prevent operation).

(3) View from outside obscured due to curtains/blinds/drapes being closed or because goods are stacked against windows.

(4) Automatic sprinkler system main valve turned off or partly closed (check that the main valve is locked in the fully open position, that the alarm gong operated satisfactorily, and that the drain valve was fully closed).

(5) Sprinkler heads nearest the fire did not operate (check operating temperature and see if heads have been tampered with to stop or delay operation).

(6) Goods stacked so high that they prevent sprinkler heads operating or restrict the area covered or poor housekeeping resulting in trails of combustible material spread throughout the premises.

(7) Recently constructed offices, screens, ceilings which have not been covered by alterations to the sprinkler system and which therefore restrict the area covered by these systems.

(8) Other fixed installations designed to work automatically did not operate.

(9) Automatic fire detection did not operate.

(10) Manual/electrical alarm not operated by occupier.

(11) Automatic dialling unit or land line to central control not operated.

(12) Fixed hose reels not used by occupier or not working.

(13) Portable fire extinguishers not used by occupier or not working.

(14) Anything which is found to be in the area and yet is obviously out of place, e.g. timing devices, flammable liquid

or gas container, broken bottles, waste rags, additional heaters (especially flame or radiant heaters) in air-conditioned or centrally-heated building.

(15) Several seats of fire or trails laid from one source.

(16) Cupboards or closets open to expose contents to flame or heat source.

Regarding the non-operation of equipment mentioned above, it should be remembered that well regulated premises keep a record of all defects in fire equipment, to whom the defect was reported, and the action taken to remedy it; and the investigator should always check into this. Even in businesses not so well regulated, it is usual for them to inform someone, if some major equipment is inoperative, viz. the local fire service or their insurance company or the installers of the equipment.

(c) Other Indications

(1) No apparent cause of fire.

(2) Fire occurred in an improbable place.

(3) Behaviour of burning material unusual.

(4) Safe door, filing cabinets, desk drawers, etc., showing signs of forced entry or attempted forced entry.

(5) Machinery physically damaged before fire.

(6) Vehicles moved from the place where they are normally kept.

(7) Animals normally within premises released.

(8) Anything which has obviously been removed from the premises, e.g. money, files, plans, pictures, insurance policies, T.V. sets, washing machines or money from slot meter, items of sentimental value (if possible an inventory of goods present after the fire should be taken for comparison with occupier's statement later).

(9) Notices in the area prohibiting smoking.

(10) Signs of carelessly discarded smoking materials. (If no smoking area, employee might have discarded cigarettes to prevent discovery and is frightened to admit it.)

(11) Limited use of ash trays in the area.

Having completed the general examination the investigator should have already reached certain conclusions, but more importantly he should also have a series of questions ready which will confirm and expand any initial theories which he may have developed.

(2) Area of Origin

Before a detailed examination of the area of origin can be undertaken, it is necessary to find the area of origin, and if the building has collapsed this can prove to be a very difficult job. It must always be remembered that the point where the fire first showed itself, or the areas of most intense burning or greatest damage, do not necessarily indicate the point of origin. However, whatever type of building is involved, the following basic facts about flame and heat travel and collapse should be borne in mind when tracing the point of origin.

(a) Fire burns upwards with far greater ease than it burns downwards. In many cases, holes are burned through floors only because some material which was burning above the floor dropped onto the floor, thereby allowing established flame to come into contact with the floor. This is a fairly frequent occurrence under or around chairs, beds or curtains that have been on fire. It also happens when a person's clothing is ignited and the person falls to the ground.

It is rare for a fire which starts at floor level to cause anything more than deep char damage to the floor. This is because the bottom layer of material protects the floor beneath it, even if the material is a flammable liquid. In fact, for a flammable liquid to cause damage to the surface on which it rests one of two things must occur; either the liquid must be completely consumed and traces soaked into the floor burn the top layer of floorboards or the liquid must leak to a lower level, be ignited, and start to burn from underneath. Even if that happened, it is by no means certain that the burning liquid on the top level would ignite that liquid which leaked through. In many cases, the surface of the material on fire which is touching the floor will not be burned, if the point of origin was at a higher level.

Therefore, a cross-section through the burn pattern would in most cases reveal very limited burning beneath the point of origin but substantial burning above it: see Fig 1. Other sur-

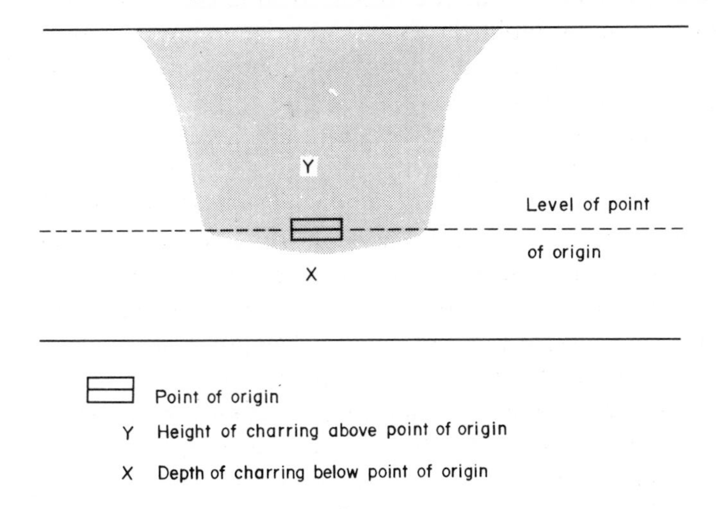

Fig. 1 Section through Point of Origin

faces facing the point of origin will obviously show a greater degree of fire damage than the reverse of those surfaces: see Fig. 2.

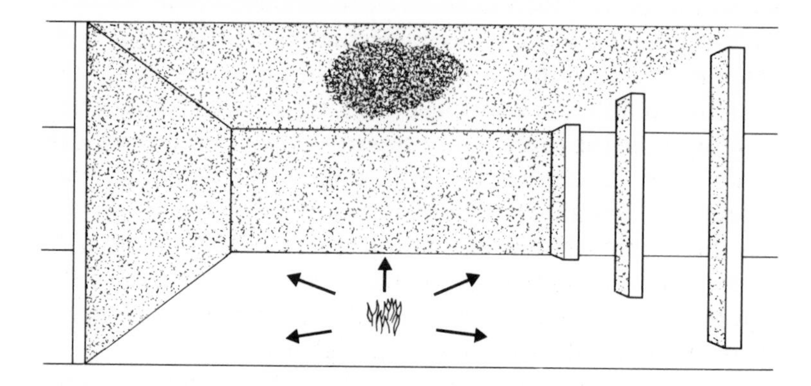

Fig. 2 Damage on Surface Facing Fire

(b) Unless the structure linings around the source of ignition can allow rapid flame spread or there are highly flammable materials laterally from that point or an exceptionally powerful airflow, the fire will not spread rapidly along a level horizontal to the point of origin.

Lateral spread occurs when the hot gasses rising from the point of origin reach a horizontal surface and spread to form a "mushroom". When this happens the burn pattern follows a vertical line from the centre of the point of origin and a horizontal line along the top of the mushroom, fading on the lower edges to form a "V" shape. The point of the "V" being the point of origin: see Fig. 3. The "V" can have very steep sides in a high building or very shallow sides when the horizontal surface is only just above the point of origin. Also the point of the "V" could be several feet wide, especially when the material at the point of origin is a flammable liquid with a large surface area.

(c) If the material burning is a liquid or gas which is leaking from a pipe or container, the flame will always burn back to the source.

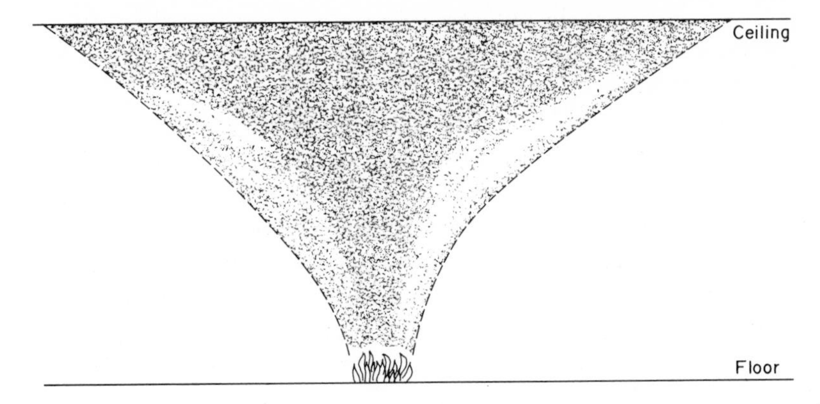

Fig. 3 The 'V' Formation

(d) Utilizing radiation, conduction and convection currents or a combination of two or more, heat sufficient to cause ignition of certain materials can be transmitted for considerable distances. It is not unknown for a fire to start in one part of a premises or even in a separate premises and, through the above-mentioned methods of heat transfer, start another fire before the first is extinguished or burns itself out. When this occurs the investigator can only find the site of the first fire by tracing all conductors and ducts and visiting buildings whose windows face the area of origin of the second fire.

Of course, with all the above basic facts the interpretation of them is a matter of degree, and in some cases it will only be possible for the investigator to determine the point of origin as being in a particular room, rather than being able to pinpoint it within a few inches.

When the exact point of origin cannot be pinpointed due to the extent of damage caused by the fire, e.g. all plaster off the walls and ceiling, the investigator should follow the procedure outlined below for collapsed buildings, *Excavation and Reconstruction.* The collapse of the plaster is treated in the same way as the collapse of a floor. The same also applies when the building was timber and it has burned down to the "foundations".

(e) *Collapse.* If a collapse occurs during a fire, the investigator must firstly determine the floor of origin. As a guide to this, the first floor to fall is invariably the floor immediately above the floor of origin of the fire. This is because of the flame and heat travel characteristics mentioned above, which result in the surface immediately above the fire collapsing because of the destruction or weakening of the supports. This will happen before any structure on the same level as, or below, the fire is substantially affected. In addition, it must always be remembered that when the original fire is in the space beneath the floor of one storey and above the ceiling of the storey below, the floor which will probably collapse first is the one containing the fire which can still be classed as the floor immediately above the floor of origin. If this is suspected, the direction of burning of the floor joists must be examined for conclusive evidence.

Depending on the structure of the fire floor or its position within the structure, e.g. concrete or timber at ground level or higher, any collapse from above will tend to cover the floor of origin of the fire. However, is some instances, the collapse of the floor immediately above the fire floor will cause a progressive collapse of other floors and could result in the floor of origin dropping to a lower level than it was at originally.

When a collapse has occurred, the physical evidence which is left depends upon the extent of the destruction. If only part of one floor has collapsed, the signs of the rate of growth of the fire and its period of burning within the area of the fire floor should be obviously different from the same signs in the area of the other floor (see Chapter 5, *Limits of Fire*). The fire floor may show a rapid or slow rate of growth, but will always show a longer period of burning than the other floor;

on the other hand, the other floor will almost always show a rapid rate of growth and a lesser period of burning than the fire floor. The investigator should exercise caution in premises having chutes, hatches, shafts or other openings in floors, as this may create the impression of a greater period of burning at the floor level where mushrooming starts rather than where the fire started. (For how openings affect and are affected by mushrooming, see Figs. 4, 5, 6.)

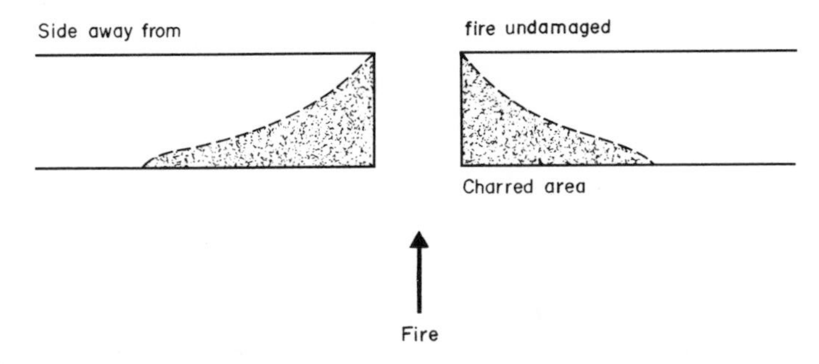

Fig. 4 The Inverted 'V' Formation (Fire Passing through a Gap)

This example holds good for a progressive collapse also, if parts of every floor remain *in situ.* In fact, if a collapse of floors results in a partial collapse of the roof as well, it usually makes it that much easier for the investigator to determine the floor of origin of the fire, because after the fire has vented very little mushrooming takes place and the damage to the area of the fire floor will be all the more obvious when compared to the other floors.

Where the collapse is in the form of a "pancake" and the parts of the floors which remain *in situ* are too small to be of any value to the investigator, he must resort to "excavation and reconstruction".

(f) *Excavation and Reconstruction.* Basically, this means that the debris needs to be examined very carefully to decide

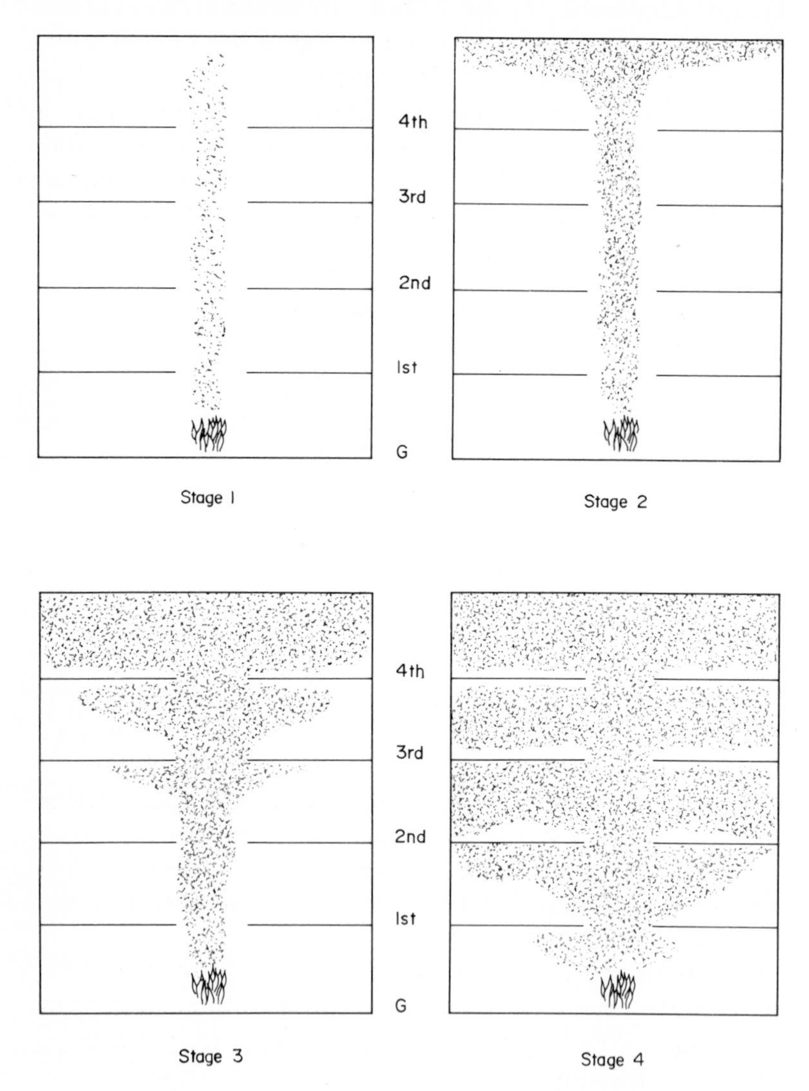

Fig. 5 'Mushrooming' (Fire under Openings)

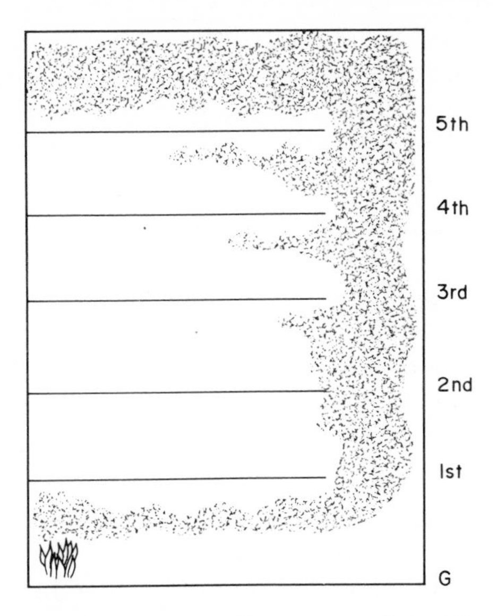

Fig. 6 'Mushrooming' (Fire remote from Openings). As with the fire under the openings, smoke density would progress downwards. However, the fire floor and the top floor will show greatest exposure to smoke and heat unlike the previous example.

the order of removal of the fallen parts of the structure, so that any evidence is not needlessly destroyed. Mechanical shovels, cranes, etc., may have to be used and dangerous parts of the structure may have to be demolished, but whatever action is taken, the investigator should make his requirements known and should be on site while work is in progress.

As the various levels are removed a visual record should be maintained by drawings or photographs, and sections of the structure and samples of the contents of each level should be taken.

The sections of the structure and the samples of the contents which have been recovered should be reconstructed on a flat surface such as a roadway or car park, so that the direction of burning of each level can be determined. It may also be possible to find out which floors collapsed through being exposed to the fire, and which fell as a result of progressive collapse.

When the lowest level at which burning has occurred is reached, clearing operations should be halted and the investigator must make an inspection of the area.

If the investigator is satisfied that he has identified the area of origin, he must then institute a thorough search of the area to find the actual point of origin and then the source of ignition and the material first ignited. If it is not possible to locate the point of origin or even the area or floor of origin, the search should be made of the remains of the premises as a whole with equal thoroughness to determine the source of ignition and the material first ignited.

(3) Determining the Source

Examples of sources of ignition and materials first ignited are given in the chapter of that title, and the investigator must check to see if any one of them was present when the fire occurred and in such a position, state and at a temperature sufficient to ignite the material which was ignited first.

If it is possible to reconstruct the contents of the room or area where the fire occurred and to pinpoint the exact point of origin, it may be that only one source of ignition and material first ignited fits the pattern of burn and other evidence.

It is probable, however, that there is more than one possible cause present as many premises contain electrical and gas appliances in the same room where smoking is allowed, etc.

Accordingly, the investigator must examine every possible cause, including remote ones such as the sun's rays, lightning, and radiation from another building; and he must try to discount them one by one, until ultimately only one remains. It is reiterated that all physical evidence, however conclusive it appears, must be considered against statements of witnesses and other documented evidence before a final conclusion is reached as to the cause of the fire.

It is hoped that the following indications and statements will assist the investigator in confirming or discounting a particular source of ignition:

(a) Electricity

Fuses—Check for correct rating and that all circuits are protected.
—If broken—blackened bridge indicates a short circuit (the fuse has burned)
—clean bridge indicates overheating (the fuse has melted)
(The investigator must check that the blackened bridge is not from a previous fault. If it is, it could mean that a higher rated fuse was installed which has allowed overheating elsewhere in the circuit.)

Switches to all appliances, motors, lights, etc.

Mains
Intermediate
Thermostatic
Time } Check whether working, and whether in ON or OFF position and look inside to check polarity.

Cable—short—Check for proximity of other conductor and signs of intense heat following a very rapid rate of rise.

Cable—overload—If the wire is loose inside the insulation, it is a sign of internal heating. If the insulation is stuck to the wire, it is a sign of external heat. However, it is possible for an external fire to strip the insulation from a part of the cable. If this happens, then conducted heat along the cable could cause the internal wire to become loose along part of the cable, but the point where the bare wire protrudes from the insulation will show signs of external heat.

When examining cable, the investigator should always find out where the cable was routed before the fire started by checking for securing clips, battens, and cable trays. If the cable was at high level and the point of origin of the fire was at a lower level, then it is reasonable to assume that any defects in the cable were caused by the action of the fire and did not start the blaze. It is important to make sure that the cable securing devices failed because of the fire, and that the cable has not been placed in position as part of a set fire. (See Chapter 3, *Arson: Timing Devices.*)

Motor—Shaft seized to bearings indicates internal burning.
—Coatings of windings melted means internal burning
—External metal of motor hotter than surrounding metal is a sign of internal burning. (Check normal temperature of motor case. Some run hot as a matter of course.)

Appliance—Check inside; if fire damaged, then fire started inside.
—Normal heat-producing appliance; check proximity of material first ignited.

(b) Gas

Type —Manufactured, Natural, L.P.G., L.N.G.

Valves—Meter
—Bulk supply
—Small containers
—Intermediate
—Thermostatic

} Check whether working and whether in ON or OFF position. Look inside, to check spindle and valve seating.

Safety Devices—Fail-safe appliance or part of appliance: operation

Leaks—Check appliances and pipework (call in specialist).

Use —Usually naked flame
—Protected combustion (internal combustion engine).

Airflow—Check collection point of air for combustion and exhaust flue.

(c) Oil-Burning Equipment

The checks necessary for oil-burning equipment are the same as those necessary for gas-burning equipment.

(d) Cigarettes

In furniture: The cigarette must be insulated by the furniture in order to build up sufficient heat to cause burning. An incubation period between 45 min and 1½ hr is necessary, and even then it will not ignite certain materials: see example in Chapter 1, *Sources of Ignition.*

A fire inside furniture is almost always caused by a cigarette, and a sign of internal burning is when it can be seen that the springs of the chair or mattress have softened and fallen, even though the fire has not spread to the remainder of the same item of furniture. (This usually happens after approximately 3 hr of smouldering.)

(e) Sun's Rays

Direction—Take compass bearing and check surrounding property to see if sun can shine through the window/roof light without obstruction.

Lens —Bull's-eye window pane
—Telescope/binoculars
—Concave mirror
—Goldfish bowl, drinking glass, glass vase

Period—Check with weather record centre, meteorological office, for periods of actual sunshine before time of origin of fire.

(f) Hot Surfaces/Naked or Open Flame

Suspected ignition by	Radiated Heat Conducted Heat Convected Heat Pyrolysis Naked Flame	Check temperature attained and period of time that temperature is sustained.

As stated at the beginning of this subsection, whatever source of ignition is eventually determined, it must be linked to the material which was ignited first, and it must be a physical possibility for the one to ignite the other.

2. Motor Vehicles

The investigation of fire in motor vehicles can prove to be more difficult than investigating property fires, particularly if the intention is to prove arson. This is because of three basic facts applicable to motor vehicles: there is a flammable liquid present in normal circumstances; the vehicle can be taken to a remote area where it is unlikely that anyone would witness the crime; and the vehicle owner can report the vehicle stolen, thereby dissociating himself from the vehicle when the fire occurs.

(1) General Examination

(a) Signs of Forced Entry

In addition to looking for signs of forced entry to the vehicle interior, the engine compartment, glove box, and luggage/load compartment must also be checked. As with buildings, the investigator must make sure that the forced entry was not made lawfully by the firemen, etc. Also, as with buildings, if there are signs of forced entry then it is reasonable to assume that malicious ignition is a probable cause of fire.

(b) Aids for Fire Growth

As stated above, flammable liquid is present in normal circumstances, and because of this the vehicle's own fuel is often the aid used to spread the fire. However, let us consider what the usual circumstances are, relating to fuel and the vehicle. Firstly, the fuel is in a container fitted with a screw top; the tank is usually remote from the engine and at a lower level than the engine fuel intake; and the only way to get fuel from the tank to the engine is by means of a pump (either mechanical or electrical) which will only operate when the ignition is switched on (electrical pump) or when the engine is being turned over (mechanical pump). Therefore, under normal circumstances, the fuel will not assist the fire, because the tank and its associated pipework always contain fuel as a liquid or vapour in such a concentration that it is above its upper flammable limit.

Accordingly, for the fuel to be involved in the fire, it must have escaped from the tank or pipework and collected in sufficient quantity to ensure a sustained fire once ignited. The investigator must, therefore, look for evidence of any leaks from the system, and he must also check all joints and pump mechanisms. If a hole is found in the tank or pipework, or joints have been loosened, or the fuel pump has been tampered with to give an increased flow, then these findings must be checked when interviewing the owner.

It is possible that fuel from a separate source to that in the vehicle tank is used, or the fuel is syphoned from the tank to aid fire growth. Indications of this would be if the fuel tank cap had been removed, a container or syphon tube found nearby, fuel on top of the vehicle body, or inside the vehicle, or fuel under the vehicle without there being a leak. An obvious check which should be made, if the fire has been aided by a flammable liquid, is to compare the liquid involved in the fire with the fuel in the vehicle's tank. (Even in a serious fire, it may be possible to obtain a sample of material from the vehicle or a sample of the soil or road surface under the vehicle for analysis and comparison with the fuel in the tank.)

(c) Other Indications

(1) Tools, other valuable objects (radio, etc.), sentimental objects removed from the vehicle. If possible an inventory of goods present after the fire should be taken for comparison with the occupier's statement later.

(2) Footprints near the car.

(3) Check any tyre prints with the tyres on the vehicle. (The section of the tyres actually touching the ground is usually untouched by fire.) Different prints could indicate that another vehicle was present or that new tyres had been exchanged for old before the fire. Keep a sample of the tyres for comparison with the owner's statement later regarding the type of tyre fitted, where bought, how many miles run, etc.

(4) Check with the residents in the area, delivery men, etc., whether the vehicle or the owner had been seen in the same area at the time of the fire or at any time previous to the fire.

(5) Check with refuelling stations near the owner's home, along the route to the fire, and in the neighbourhood surrounding the fire, whether the vehicle owner had purchased fuel, and if so whether it had been supplied in a separate container.

(6) Check whether the owner normally travels along a route which takes him near the spot where the car caught fire.

(2) Area of Origin

Because of the size and shape of most vehicles, the "V" pattern of burning is not usually obvious. However, the fire will still have a recognizable pattern for the investigator to trace to the point of origin. For example, the fire will start in one of four places—in the engine compartment; in the driver/passenger compartment; in the luggage/load compartment; or outside. If the fire starts in more than one place, it is usually an indication that a flammable liquid has contributed to the fire, and the investigator should follow the checks detailed in the above two sections. (He should not, however, overlook the possibility of an electrical fault setting fire to the wiring harness and thereby spreading throughout the car.)

If the fire has started in only one of the places mentioned above, then it will follow a path compatible with conduction, radiation and convection currents, i.e. from the front of the vehicle to the rear, from the tyre up the side of the vehicle, or from the driver/passenger compartment out through the windows.

Because of the protection afforded them by their postion or surroundings, the engine mountings, the fan belt, and the lower part of the tyres are not generally affected by fire. If they are, a careful check should be made for traces of flammable liquid. It must always be remembered though that there are often oil deposits around the engine and on the underside of a vehicle which may assist a fire. Also, exterior paintwork can be scorched or catch fire by conducted heat from a fire inside.

(3) Determining the Source

The sources of ignition for vehicle fires are much the same as those indicated for fires in buildings, though not, of course, as numerous; and as with a building fire, the investigator must discount the possible causes one by one until only one remains.

(a) Electricity

Fuses—Check for correct rating and that all circuits are protected, including any auxiliary circuits for extra lights, radio, etc.
Switches—Check whether on or off, and check polarity.
Cable—Check for short or overload, as for buildings.

(b) Cigarettes

Check, as for buildings. (Remember that it is difficult, if not impossible, for a cigarette to ignite petroleum spirit.)

(c) Sun's Rays

Check as for buildings; but remember that the type of windows fitted to vehicles vary widely, and many restrict the power of the sun's rays. Also this cause only holds good for a fire in the driver/passenger compartment.

(d) Hot Surfaces

Usually the only surface in a vehicle which is hot enough to cause ignition is the exhaust manifold.

(e) Backfire

If the engine valves are not working correctly, it is possible for either the inlet or exhaust valve to lift when the cyclinder fires. When this happens, flame passes along the inlet pipe or the exhaust pipe. However, the only time that this flame will cause a sustained fire is when there is a break in the exhaust system, or the carburettor is not fitted with an air-cleaner element (the element acts as a flame trap), and the flame is allowed to reach some combustible material.

(f) Runaway

One unusual cause of fire occurs only with diesel engines which are driven into a gas cloud of certain flammable substances or alternatively are running while stationary and are subsequently surrounded by such a cloud. What happens is that the gas is entrained into the engine air intake and the engine runs faster and faster until it explodes.

As with building fires it is essential to link the source of ignition to the material first ignited.

Chapter Ten *The Investigation: Interviewing; Documented Evidence*

THE COLLECTION of information is quite separate to the collection of physical evidence in that it generally takes place outside the building which was involved in fire and can involve considerable time in travelling and then searching out the correct witnesses or documents. It is possible, of course, for some of the information required to be obtained by telephone, but if at a later date the evidence has to be presented in a court of law, the investigator must obtain a written, signed statement (a telephone conversation or indeed any conversation related to a court by the investigator could be classed as hearsay and as such would not be accepted as evidence).

Where documents are concerned, the investigator should make every attempt to obtain an authenticated photostat copy, even though the court may have the power to demand sight of the original document later.

The information required falls into a natural time scale, viz. previous history, immediately prior to the fire, during the fire and after the fire, and these headings are amplified below.

1. Previous History

(1) The Company and Its Directors

(a) Any previous fires in the same premises.
(b) Any previous fires in the same company, but different premises.
(c) Any previous fires in premises owned by the same company, but not occupied by them.
(d) Any previous fires in a different company, but with the same directors.

(2) People

(a) Any employees been associated with previous fires in this or other premises.
(b) Any former employees resident in the area who were associated with previous fires.
(c) Any residents in the area who have been associated with malicious ignition.

(3) Structure

(a) Obtain former building plans and compare them with the remains of the premises after the fire, to check for any structural alterations.

(b) Check the installation of false floors, walls and ceilings.

(c) Check former plans for built-in chimneys, ducts, shafts and other cavities which may have assisted fire spread, if they were not sealed properly or the seal was combustible.

(d) Obtain copies of plans/reports (requirements and recommendations) made by:
   (1) Fire Service
   (2) Loss Control Company
   (3) Risk Manager
   (4) Insurance Company
   (5) Sprinkler Engineers

and check the amount of work carried out. Also check for alterations made since any such work was carried out, which might have reduced the standard of protection previously achieved.

(4) Values

(a) Check the current value and insured value of the *structure,* who benefits and by how much.

(b) Check the current value and insured value of the *contents,* who benefits and by how much.

(c) Check the insurance policies:
   (1) Date policies taken out.
   (2) Dates values increased and by how much.
   (3) Is the policy index linked or a "valued" policy?
   (4) How many policies are in force for the premises, and which insurance companies have an interest?
   (5) What types of cover are provided (direct loss, consequential loss, etc.)?

(d) Check the volume and value of trade in the business involved; look for downward trend.

(e) Check the credit rating of the company/occupier/owner, to see if bankruptcy was a possibility.

(5) Hazardous Substances

(a) Was there adequate structural protection provided?
(b) Did all flammable liquid tanks have a overfill alarm or some other safety device fitted?
(c) Was the manufacturer's recommended storage code complied with (right way up, out of sunlight, away from heat, grease, etc.)?
(d) Were incompatible substances stored together?
(e) What cylinders or pressure vessels were within the premises?

(6) Services/Fire Protection Devices

(a) Is there a history of any defects or malfunction in the heating system, lighting circuits, power supply, friction machinery (belt drives, overloads), sparking machinery or machinery using naked flames or other heat which may have caused the fire?
(b) Is there a history of any defects of malfunctions in the fire-protection devices?
(c) What form did the defect take?
(d) If defects have occurred, were they persistent or intermittent?
(e) How often had the defects been reported and to whom?
(f) What steps had been taken to rectify the problems?
(g) How much time has elapsed since the last defect ocurred?

(h) Was there a realistic programme of inspection, routine maintenance and major overhaul for services and fire-protection devices.
(i) Was the programme adhered to and a record maintained?

(7) Practices Carried On

(a) If the type of work done involves bonus or piecework, have "short cuts" developed which could affect safety?
(b) What is the general standard of housekeeping in the premises?
(c) How does the company dispose of waste material?
(d) If a no smoking rule is imposed:
    (1) Are smoking areas provided?
    (2) Are employees allowed to carry matches and lighters?
    (3) What level of supervision is maintained?
    (4) What are the penalties for breaking the no smoking rule?

2. Immediately Prior to the Fire

(1) The Directors

(a) Where were they, when the fire started?
(b) Where were they, when they were informed of the fire?
(c) What was their reaction on being informed of the fire?
(d) Was there anything unusual about their dress when informed of the fire (e.g. fully dressed and awake in the middle of the night)?
(e) Did they arrive at the fire much quicker than would have been possible? (Was someone watching from nearby to check that the fire had a good hold?)

(2) People

(a) Was anyone smoking in a no smoking area?
(b) Was anyone carrying matches in a restricted area?
(c) Had any visitors been in the fire area, e.g.
 (1) delivery men,
 (2) maintenance/repair men,
 (3) outside contractors,
 (4) workers from other departments,
 (5) friends of employees,
 (6) family of employees,
 (7) former workers?
(d) Who was the last to leave the fire area, and what time did he leave?
(e) Was anyone seen leaving the fire area or the building immediately before the fire was discovered?
(f) In the days prior to the fire, had any employees been
 (1) admonished,
 (2) warned,
 (3) cautioned,
 (4) suspended,
 (5) given notice to quit,
 (6) dismissed,
 (7) laid off,
 or displayed any ill-feeling or dissatisfaction with the company, his job, fellow workers, etc.?

(3) Contents

(a) Check the normal parking area for vehicles, and compare with the physical evidence, to see if vehicles had been parked away from the premises.
(b) Compare the remains of the fire area with the former layout and/or plant location plans.

(c) Obtain an inventory of contents, machinery, furnishing, fittings, raw material, material in process, finished goods stored, and compare it with the remains in the premises. Also check if it would have been physically possible for the quantity of goods claimed on the inventory to have been housed in the premises.

(d) Check with neighbours, police patrol, security patrol, if more goods than usual had been removed from the premises, and if any process machinery had been removed from the premises.

(e) If goods have been removed, then obtain a description of the goods and ascertain their value.

(f) Also check the above with any haulage companies seen in the premises or moving goods from the premises.

(g) Are animals normally kept on the premises? If so, compare with the physical evidence to see if they had been released before the fire.

(h) Any change of use of the premises taken place since the premises were last inspected by the organisations mentioned at (3) (d) in the subsection, *Previous History.*

(i) A check should be made on when the premises were last occupied. If not recently vacated, check with delivery companies, meter readers, etc.

(4) Services

(a) Electrical Installation

Were the switches controlling the following, "on" or "off" when the fire occurred? If they were "on" when the fire started, did anyone then switch them "off", and if so in what sequence? (In some cases, the normal cut-off switch for a machine can isolate the manufacturing capability of the machine, but keep heating, ventilation or transportation systems in operation.)

(1) Main intake
(2) Distribution box
(3) Power circuits
(4) Lighting circuits
(5) Ventilation circuits
(6) Individual motors
(7) Appliances (television, refrigerator, etc.)

(b) Gas Installation/Oil Installation.

As with the electrical installation, the question of whether the gas valve was "on" or "off", etc., must be determined for the following controls:

(1) Main valve
(2) Intermediate valve
(3) Valve to individual appliances

(c) Machinery

(1) Did the machinery working at the the time of the fire exhibit naked or open flames or sparks, or was heat caused through friction or overloads or from any other cause?
(2) Were all cooling systems working satisfactorily?

(5) Fire Protection

(a) Was the view through windows from outside obscured?
(b) Were windows open or shut?
(c) Were doors open or shut (in particular FIRE doors)?

(d) Were all fire-protection systems ready for operation?
(e) Had there been another fire in another part of the premises or in adjoining or nearby premises?

(6) Practices Carried On

(a) Were any malpractices taking place with regard to work routines?
(b) Were any new systems or new equipment being evaluated?
(c) Was there a lot of waste material in the fire area?
(d) Was the waste material in the process of being disposed of?
(f) Had a supervisor or security man recently checked the area?
(If someone is smoking in a restricted area, he may dispose of his cigarette hurriedly and carelessly, if he thinks that he may be discovered.)

3. During the Fire

Questioning Witnesses

This subsection is presented as a *pro forma* for an interview sheet, rather than as a straight list; and it can be used to question any witness whether from within the premises involved (employee, security, management, owner) or from outside (fireman, policeman, passer-by, neighbour). It is, of course, essential for the person discovering the fire or first arriving at the scene to be interviewed.

It must be borne in mind, however, that a witness's statement may not be as accurate as the investigator would wish. It might be misleading or deliberately false—an over amplification of the facts or a composite of what the witness has seen and what he has heard others talking about. There is a certain skill in this type of interviewing, particularly when dealing with small children and distraught witnesses, and the investigator should constantly endeavour to improve his interview technique.

Not all of the questions will apply in every case, and in some instances additional questions will be necessary. As stated previously, the object of collecting information is to weigh it against the physical evidence, in order to determine as accurately as possible the cause of the fire.

(a) The Witness

(1) Name
(2) Address
(3) Relationship to premises
(4) Time of arrival/discovery of fire
(5) How did you become aware of fire?
(6) What were you doing when you became aware of the fire?
(7) Where were you standing when you became aware of the fire?
(8) What time did you arrive at the fire?
(9) When was the Fire Service called?
(10) What did you do?
(11) What did you see?
(12) What do you think caused the fire?
(13) Did you think that there would be a fire one day? Why?
(14) Has anything been moved out of the premises?
(15) Has anything been moved out of its usual place?
(16) Is there anything extra in the premises?
(17) Who could have had anything to do with the fire?

(18) Who else might know something about the fire or premises?

(b) The Fire

(Check with different witnesses—there may be more than one seat of fire)

(1) Where was the first smoke seen? What colour was it?
(2) Where was the first flame seen? What colour was it?
(3) Was there an unusual smell immediately before the smoke or flame was seen?
(4) Was there a distinctive smell during the fire?
(5) When did lateral and/or vertical spread occur (time scale)?
(6) How quickly did the spread occur?
(7) Did the character of the smoke change? When? How?
(8) Did the character of the flame change? When? How?
(9) Were there any sharp noises/explosions? When? Where?
(10) Was there any collapse? When? Where?

(c) Fire Protection/Fire Fighting

(1) Did any automatic fire-detection system operate?
(2) Did any automatic fire-attack system operate?
(3) Was a fire-alarm system operated manually?
(4) Did anyone attack the fire with fixed hose reels or portable extinguishers?

(d) Structure

(1) Were any doors open or broken?
(2) Were any windows open or broken?
(3) Was there any sign of forced entry through hatches, cellar flaps, manholes, roofs, etc.?

(e) People from the Premises

(1) What was their attitude?
(2) What were they dressed in (night clothes/indoor clothes/outdoor clothes)?
(3) Was anyone rescued? What was his name? Where is he now?
(4) Were there any casualties? What were their names? Where are they now?

(f) People Nearby

(1) Anyone seen running away from the premises.
(2) Did any people watching the fire give rise to suspicion because of their attitude, e.g.
    (a) Hiding from view;
    (b) Unusually excited (particularly in a sexual way);
    (c) Anticipating spread;
    (d) Obstruction of firemen?
(3) Anyone seen at previous fires.
(4) Who was the rescuer? Where did he come from? Where is he now?

(g) Weather

(All measurements required as general values, e.g. hot/cold. If more detailed information is required, it can be obtained from the weather centre.)
(1) What was the temperature and humidity?
(2) What was the wind strength and direction?
(3) Was visibility good?
(4) Were weather conditions wet/dry/raining/sleet/snow?

4. After the Fire

Insurance Claim

The investigator's principal interest in the insurance claim is whether or not the claim is realistic. To determine this, the investigator must rely not only on the physical evidence, but also on the claim submitted by the insured or his assessor and any counter claim made by the insurers or loss adjusters. In addition, the view of the Fire Service as to the amount of damage should be sought, and also as to the amount of salvage work carried out by them.

A check should also be made on the quantity of material and plant which has suffered damage in the fire (whether from flame, heat, smoke, or water) but which can be reclaimed or refurbished.

5. Conclusion of the Investigation

On completion of the investigation it is necessary for the investigator to collate carefully all of his notes, drawings, photographs, samples and reports from specialists (chemists, pathologist, etc.). From all of this information, he should prepare a concise report stating briefly the possibilities which he considered and his reasons for reaching the conclusions stated.

All of the items referred to above should be kept for reference purposes, at least until any court cases or insurance claims have been finalised.

Chapter Eleven *Abridged Case Histories of the Six Categories of Arson*

(a) To Gain Financially

THE OWNER of a clothing factory cum warehouse in financial difficulties did not discourage bad housekeeping practices amongst his workforce, with the result that cardboard boxes adjacent to each sewing machine were overflowing with off-cuts and rejects. All windows had been blanked off with corrugated carboard to "stop the sunlight from fading the garments". Finished garments were hung on free standing moveable rails (about 20 dresses per rail) throughout the work floor.

On the night of the fire, the owner did not leave his premises until after 9.00 p.m., when all other businesses in the area had been closed for almost three hours and darkness was beginning to fall. Unfortunately for him, he had forgotten about a skylight covered with thick black curtaining. The curtaining burned and the flames in the skylight were spotted by the caretaker of a nearby office block, only minutes after the owner had left.

After the fire had been extinguished, a two kilowatt electric fire was found with its elements still on, adjacent to one of the free standing moveable rails. It was later ascertained from workers that the electric fire was normally in the small office used by the owner. Tests carried out by a research laboratory showed that the garments were easily ignited by radiated heat and once ignited burned with great rapidity.

(b) To Conceal Another Crime

A shop assistant/delivery-van driver decided to steal money from the safe in the shop where he worked. He returned to the premises late at night, entered by forcing the lock on one of three rear doors, and made a very amateurish attempt at safebreaking. He did not succeed in opening the safe but did leave cut marks on the safe door. He then made a bonfire in the centre of the premises, using any combustible goods which came to hand, lit the bonfire and left the premises by the same door through which he had entered. The fire was very severe, involving the ground and first floor of the premises and eventually breaking through the roof.

Suspicion was aroused initially, because even though the fire was so severe it was possible to pinpoint the point of origin, but a physical inspection of the point of origin and statements of witnesses failed to reveal anything which could have acted as a source of ignition. It was then discovered that the fire crews who attended the fire had only broken open two doors at the rear of the premises, but all three doors showed signs of forced entry. Accepting the investigator's findings, the police made local enquiries which led to the arrest of the shop assistant.

(c) To Destroy/Protest

Late one evening, a fire occurred at a day attendance centre for "problem" children. The fire was discovered soon after ignition by a man who was taking his dog for a walk; and the subsequent early attendance and attack by the Fire Service resulted in the fire being extinguished before serious damage was caused.

In the room of origin, books, desks and chairs had been piled in a heap in the centre of the floor and ignited. A large kitchen within the premises had been ransacked, and bags of flour had been torn open and the contents scattered over all working surfaces and the floor. Children's footprints were found in the flour; and a nearby children's home had reported to the police that two of their charges (who were also pupils at the day centre) were absent. The children were found by the police and admitted breaking into the day centre and starting the fire.

An interesting development in this case was that, although the pupils at the day centre came from a variety of different locations (and it was not possible to inform them not to attend at 0800 hours the next day), none of them turned up the morning after the fire, which indicates that they all had prior knowledge of the incident.

(d) The Hero

A policeman posted to a quiet rural area became dissatisfied when he saw various colleagues getting promoted or having a part in large scale investigations.

After his wife divorced him, he started to "discover" fires in barns and other farm property and received praise from his senior officers. However, suspicions were aroused when the policeman reported a fire quicker than he could have done under normal circumstances. A check on recent fires in the area resulted in the policeman being questioned, charged and convicted of arson.

(e) Mental Disorder

While controlling a crowd of onlookers watching a spectacular warehouse blaze, a policeman's attention was drawn to a young man who seemed to be much more excited and more mobile than the other spectators. On closer inspection, it was found that the man was sexually aroused; and he was questioned about the cause of the fire. He subsequently admitted setting fire to the warehouse and claimed responsibility for two other fires in the area.

(f) Boredom

A man carrying out a dull, repetitive work routine at a military installation, lit a fire to cause some excitement and the resulting blaze claimed the lives of two firemen. The man, overcome with remorse, admitted starting the fire.

APPENDIX: Abridged Case Histories of Causes of Fire Other than Arson

(a) Pyrolysis

In a large factory, all of the machines were driven by their own electric motors via M.I.C.C. cable. Each distribution box controlled twelve of these motors and the cable to them was routed between the timber boards of the double thickness floor. For a short distance, all twelve cables ran in a common trough before branching off to their respective machines.

Over the years, larger, more powerful, machinery had been installed utilizing the same power supply, and as product demand increased so the machines ran for longer periods.

The result was a build-up of heat in the trough which culminated in pyrolysis of the timber floorboarding.

(b) Sun's Rays

After having Georgian-style windows installed (some of the small panes being the "bull's-eye" design) the owners went on holiday for two weeks. On their return they discovered scorch marks along the back of a simulated leather settee. Investigations showed that the focal length of the lens formed by the "bull's-eye" glass coincided with the location of the scorch marks.

(c) Candle

A divorcee with three small children could not pay the electricity bill in the multi-storey block of flats (apartments) where they lived, and the supply was cut off. As the entrance hall, toilet and bathroom did not have any natural illumination the occupier fixed candles in those areas.

The candle in the toilet was placed directly on the lid of the low level plastic cistern and one of the children left the candle alight after leaving the toilet. As the candle burned down, it set fire to the cistern lid, which subsequently spread the fire to the plastic-tiled walls and ceiling and into the hallway.

(d) Electrical

Four terraced houses were converted into a night club and the interior was extensively altered and false walls and ceilings installed. After a disastrous fire occurred one morning, it was found that only one of the former four electrical intakes was "official". The intake which had provided the source of ignition (by overheating) did not enter the premises via fuse boxes and meter. Instead, several leads to various electrical appliances had been fed into a copper tube. The end of the copper tube had then been hammered flat to secure the leads and the other (open) end of the tube had been pushed over the main cable intake.

Index